Success and Creativity in Scientific Research

Success and Creativity in Scientific Research

Amaze Your Friends and Surprise Yourself

David S. Sholl

CRC Press
Taylor & Francis Group
Boca Raton London New York

CRC Press is an imprint of the
Taylor & Francis Group, an **informa** business

First edition published 2021
by CRC Press
6000 Broken Sound Parkway NW, Suite 300, Boca Raton, FL 33487-2742

and by CRC Press
2 Park Square, Milton Park, Abingdon, Oxon, OX14 4RN

© 2021 Taylor & Francis Group, LLC

CRC Press is an imprint of Taylor & Francis Group, LLC

Library of Congress Cataloging-in-Publication Data
Names: Sholl, David S., author.
Title: Success and creativity in scientific research : amaze your friends
and surprise yourself / David S. Sholl.
Description: First edition. | Boca Raton, FL: CRC Press, 2021. | Includes bibliographical references and index. | Summary: "Long-term success in scientific research requires skills that go well beyond technical prowess. This book is aimed at students and early career professionals who want to achieve the satisfaction of performing creative and impactful research in any area of science or engineering. Both entertaining and thought-provoking, this essential work supports advanced students and early career professionals across a variety of technical disciplines to thrive as successful and innovative researchers"—Provided by publisher.
Identifiers: LCCN 2020043700 (print) | LCCN 2020043701 (ebook) |
ISBN 9780367619190 (hardback) | ISBN 9780367619183 (paperback) | ISBN 9781003107095 (ebook)
Subjects: LCSH: Research. | Research—Methodology.
Classification: LCC Q180.5 .S56 2021 (print) | LCC Q180.5 (ebook) | DDC 507.2/1—dc23
LC record available at https://lccn.loc.gov/2020043700
LC ebook record available at https://lccn.loc.gov/2020043701

ISBN: 978-0-367-61919-0 (hbk)
ISBN: 978-0-367-61918-3 (pbk)
ISBN: 978-1-003-10709-5 (ebk)

Typeset in Palatino
by codeMantra

Contents

Preface

This book aims to be useful for anyone who is pursuing a career involving creativity and research in a technical discipline. This broad description includes not only readers who are (or aspire to be) researchers at universities, but also a wide range of career paths in industrial R&D, startup companies, and non-profit and government organizations. When aiming to develop a career of this type, it is natural to wonder what characteristics separate the performance of people who are average from the exceptional. Technical skill and knowledge are a prerequisite for success in a research career, but they are far from being the only skills that are important. This book is about all those other skills.

The book's title and many of the chapter titles draw tongue-in-cheek inspiration from the enormous collection of self-help literature that promises to lead readers to a better life. Authors in this genre seem to engage in an arms race seeking titles that imply dramatic outcomes with less and less effort, for example, *The Stress Free You: How to Live Stress Free and Feel Great Everyday, Starting Today*. Who wouldn't want to do that? Or for those who don't want to wait an entire day for satisfaction, *Happiness Habits: Instant Happiness in 15 Minutes or Less*. This book does not offer 15-minute solutions or one-day action plans. Long-term satisfaction and success in a research career requires habits that are built up and applied over extended periods of time. The chapter titles should be viewed in the same sense as the annual Ig-Nobel Prizes, as something that "first make people laugh, then make them think".

It would be wildly presumptuous to claim that I have deep insights into career success that have not occurred to others in the past. Accordingly, much of the book's content relies on excellent books by others, including books about creativity and work habits from a diverse array of disciplines and scientific biographies. Readers could, of course, skip this volume and directly read the 20 or so books that underlie its content, but in my experience, people developing their careers are busy enough that this course of action is unlikely. My hope is that by bringing together a set of ideas from these sources. This book will motivate readers to read further in the topics that are most meaningful to them.

This book originated in an annual series of talks I have given to PhD students, postdoctoral researchers and faculty in the School of Chemical & Biomolecular Engineering (ChBE) at the Georgia Institute of Technology, Atlanta, Georgia, USA. This is a group of enormously successful researchers who are making advances at the cutting edge of their discipline. I appreciate the forbearance of my faculty colleagues who allowed me to take over a slot in our seminar series for an unorthodox use and the encouragement I have received from colleagues at all levels in the School that these talks were entertaining and useful. I also benefited from the efforts of a large group of PhD students and postdocs in ChBE who anonymously reviewed draft chapters of this book and the work Lori Federico did to organize this review process.

How to Use This Book

This book will only be helpful to you as a creative researcher if it leads to changes in the ways you approach your work. Each chapter ends with a brief summary of the chapter's main points. Upon reaching a chapter summary, I encourage you to thoughtfully consider specific changes in your habits that would be worthwhile and chart some specific steps you will take to make these changes.

Many of the ideas in the book are illustrated by anecdotes from successful scientists and other luminaries. It can be tempting to confuse correlation and causation when hearing stories like these. Sometimes, this distinction is obvious. If you read that "Prof. X, who famously wore light green socks every day of her working life, later won the Nobel Prize", I hope you don't think that making a similar choice about your footwear will lead to career success. There are many places in the book, however, where I use generalizations to describe some recommended approach to a problem. To take one example among many, you will read in Chapter 5 that prolific writers schedule their writing time. As a careful, detail-oriented researcher, you could reasonably object that there might be many examples of people who do this, but surely not every successful writer in history has taken this approach. I ask that you excuse from my writing the absence of caveats and prevarications that are necessary in rigorous scientific reports and instead understand that in these generalizations I mean that *on average* the ideas that I recommend improve professional productivity.

I hope that you find the anecdotes and ideas in the following pages inspiring and challenging. Most of all, I hope that they help you to be more effective in pursuing the creative endeavors that lie in your future.

Author

David S. Sholl is the John F. Brock III School Chair of Chemical & Biomolecular Engineering (ChBE) at the Georgia Institute of Technology, Atlanta, Georgia, USA. During an academic career of more than 20 years, he has advised more than 60 PhD students and postdocs. His research group has published more than 350 papers describing their work on computational material modeling for applications in the chemical industry and energy sector. Sholl has given more than 200 invited seminar and talks around the world and written two previous books, *Density Functional Theory: A Practical Introduction* (with Janice Steckel) and *Polyphony*, a novel. He has also participated in writing documents that defined research agendas for the US Department of Energy and the National Academies of Science, Engineering and Medicine and for 10 years was a Senior Editor of the American Chemical Society Journal *Langmuir*.

For the past seven years, Sholl has led ChBE at Georgia Institute of Technology, Atlanta, GA, USA, the largest academic chemical engineering department in the USA, with 900 undergraduate students and 250 PhD students and postdoctoral researchers. ChBE has recently been ranked as the #2 best undergraduate chemical engineering program in the USA (*U.S. News and World Report*), the #5 best graduate chemical engineering in the USA (*U.S. News and World Report*) and the #6 chemical engineering department in the world (*Shanghai Ranking Consultancy Academic Rankings*). During his time as School Chair, Sholl's research group has published more than 130 papers. This book is based on lectures Sholl has given annually to the graduate students, postdoctoral researchers and faculty in ChBE.

1

How to Become an Overnight Success

Success in short track speed skating requires high levels of technical skill and incredible athleticism. In February 2002, the Olympic final of men's short track speed skating featured several bona fide stars, including Apolo Ohno, who was favored to win a gold medal skating for the USA. The race also included a 28-year-old Australian called Stephen Bradbury who was essentially unknown outside of speed skating. As the racers came into the final turn, Bradbury was well behind the leaders and out of medal contention. In the blink of an eye, one of the lead skaters fell and slid off the track, taking the rest of the pack with him in a tangle of skates and limbs. Bradbury was the last man standing and he coasted the last meters to the finish line, his hands above his head in delight and disbelief. He had won an Olympic gold medal.

Bradbury's victory wasn't just an unlikely sporting event; it was the first time an athlete from the Southern Hemisphere had won a gold medal in any event at the Winter Olympics. Instead of simply being an item of sports trivia, the nature of Bradbury's triumph was a "you have to see this" moment that transfixed people around the globe. Overnight, Stephen Bradbury went from being unknown in his sports-mad homeland to being a national hero.

In many areas of science and culture, the longest lasting form of fame is to have your name associated with an important phenomenon or discovery. Think of Schrödinger's cat, Fermat's Last Theorem or the Krebs cycle. Stephen Bradbury's Olympic medal brought him this kind of linguistic immortality, at least in Australia, where the phrase "doing a Bradbury" is now used to describe someone achieving amazing and unexpected success. I suspect that most people who are entering a career of scientific research or similar creative pursuits quietly dream of "doing a Bradbury". Perhaps this would mean overturning a long-established scientific consensus or curing a dreaded disease or unlocking one of nature's secrets that will change how people view the world forever.

This chapter looks at a deceptively simple question: what does it take to become an overnight success in the world of research? Among others, we will meet a Nobel Prize winner, a doctor who developed a world-changing therapy and a college dropout who founded a company valued at over $10 billion. There is no "magic formula" for achieving scientific fame and fortune, but the commonalities and contrasts between these impressive individuals point towards some truths about achieving spectacular success in science that can greatly help you in your career.

Then and Now – Marco Polo, Alfred Russel Wallace and Grumpy Cat

Spending time thinking about history is a powerful way to appreciate life in the modern world, with innovations like indoor plumbing, drinkable water and reliable heating. It is also striking just how long it took for things to happen in the past. The life of Marco Polo is a dramatic example of this observation. In 1271, Marco Polo left the city of Venice with his father and uncle to travel and trade in what is now known as central Asia and China. Although these regions were already inhabited by sophisticated cultures, they were entirely unknown to Europeans. After almost two and half decades of continual travel, he returned to Venice for the first time, presumably looking forward to telling stories about his adventures and enjoying his riches. Unfortunately for Marco Polo, his city was in the midst of a war. In the year after he returned home, he was taken prisoner in Genoa, where he was held for three years. During this time, Marco Polo dictated a book about his life that we now know in English translation as *The Travels of Marco Polo*. In 1299, 28 years after he first left Venice, Marco Polo was released from prison and his book was published. In today's world, we could imagine that the book would rise rapidly up bestseller lists and propel its author to international fame. In a sense, this did happen, but much more slowly. It wasn't until well into the 1300s that *The Travels of Marco Polo* became widely available and read, spurring interest in trade routes between Europe and the Far East and ultimately making Marco Polo a household name. In 2014, Netflix spent $200 million dollars to make a series called simply *Marco Polo*. Critics savaged the series, and it seems that audiences weren't too thrilled either, because the series was cancelled after its second season. But ask yourself, how much money is someone going to spend in the future to make a fictionalized TV series based on your life story?

To find a parallel with Marco Polo's adventures in the history of science, it is hard to look beyond the English naturalist Alfred Russel Wallace. Wallace was born in 1823, a time when becoming a scientist didn't just mean being white and male ("qualifications" that Wallace met) but also required being born into a wealthy family. Wallace came from a modest background. He determined that the only way to pursue his intellectual interests was to travel to distant lands and collect exotic specimens he could then sell. This plan was put into action in 1848 when the 25-year-old Wallace sailed to Brazil to collect samples and sketch animals and plants.

After four years of collecting and sketching he boxed up everything he had accumulated and boarded a ship for the long voyage home. Twenty-six days into the journey disaster struck when the boat caught fire. Wallace had to watch from a lifeboat as his hard-won collection burned and then sank. His efforts had been for naught, and Wallace and the ship's crew were lucky to escape alive.

In 1854, Wallace left the shores of England again, this time traveling to the Malay Archipelago. He spent the next eight years exploring what is now Malaysia, Indonesia and Papua New Guinea. Aided by as many as 100 assistants from the local population, he amassed a truly astonishing collection of more than 80,000 beetles and more than 40,000 other specimens, many from species that had never been recorded by scientists before. This means that he added, on average, more than 25 beetles every day for eight years to his collection. Most people I know would be happy to see fewer than 25 beetles in a year, so this devotion to cataloging the world of beetles is quite impressive.

In the 12 years since Wallace first left England, he hadn't simply been collecting beetles. He had also been thinking deeply about what he had seen. He identified what is now known as the Wallace Line, a geographic boundary that threads through the Malay Archipelago separating regions populated by completely different groups of animals. Some observations of this phenomenon had been made as early as the 1520s by the Italian explorer Antonio Pigafetta, but Wallace made the demarcation of the line more precise and, more importantly, came up with an explanation for its existence. Wallace independently developed the theory of evolution, which was being worked on at the same time by Charles Darwin. Wallace's early writing on his ideas prompted Darwin to finish and publish his own *On the Origin of Species*, one of the most famous scientific books of all time. Wallace's extensive field observations and his origination of what is now called biogeography made him an intellectual leader in the 19th-century science.

Remarkably, Wallace's work was part of not one but two scientific revolutions. Shortly after Wallace's death in 1911, Alfred Wegener proposed that the existence of identical fossil species on opposite sides of the Atlantic could have resulted from movement of continents over vast spans of time. The biogeographical boundaries that Wallace had established could also be explained with the same ideas. For many years, Wegener's concept was considered to be speculative, if not heretical, by most geologists. In the 1960s, careful measurements of ocean floors provided direct evidence that Wegener was right – the continents separated by the Atlantic Ocean were slowing drifting apart. Today, the role of evolution in the characteristics of life on Earth and the role of plate tectonics in shaping the continents we live on are foundations of modern science.

The Internet-fueled, social media-saturated world we live in today makes the decades that Marco Polo and Alfred Russel Wallace spent in obscurity seem labored and tedious. The dour expression of the feline known as

Grumpy Cat is a prime example. Five months after her birth in 2012, a picture of Grumpy Cat's now famous downturned mouth was posted on Reddit, where it rapidly enraptured fans around the world. Six months after first appearing on Reddit, Grumpy Cat attended the South by Southwest festival in Austin, Texas, where she was reportedly more popular with crowds than luminaries such as Elon Musk, founder of Tesla and SpaceX. Months later, a book "written" by Grumpy Cat appeared, debuting at number 8 on the *Publishers Weekly* hardcover list. Publicity for the book included an "interview" in *Forbes* magazine. In 2014, Grumpy Cat cemented her status as a cultural icon by appearing on the season finale of the popular TV show *The Bachelorette*. When Grumpy Cat passed away in 2019, her death made headlines around the world.

Grumpy Cat is just one of many examples of what might be called fame at the speed of the Internet. In 2015 the parents of an enthusiastic young boy called Ryan began posting videos of him unwrapping toys on YouTube. If you are not closely related to someone under the age of seven, it is very unlikely you have ever watched one of Ryan's videos or even heard of him. But by 2017, Ryan's YouTube channel, Ryan ToysReview, was one of YouTube's most watched channels for months in a row. As of this writing, the video of 3-year-old Ryan imaginatively titled "GIANT Lightning McQueen Egg Surprise with 100+ Disney Cars" has been watched almost a billion times.

Most would agree that the "accomplishments" of Grumpy Cat or Ryan ToysReview are not analogous to world-changing scientific innovations. But the ubiquity of viral videos and other trappings of Internet culture can contribute to the sense that making a big splash in the world shouldn't take years of anonymous labor. This is not a new idea in scientific culture. A longstanding trope in mathematics and theoretical physics is that the ability of "geniuses" to make breakthroughs diminishes rapidly after the age of 30. Paul Dirac, who won the Nobel Prize in Physics in 1933 at the age of 31, captured this idea in a gloomy verse:

> Age is, of course, a fever chill
> that every physicist must fear.
> He's better dead than living still
> when once he's past his thirtieth year.

A modern day Dirac would probably be advised to use more gender inclusive language, or perhaps more prudently to simply skip poetry and stick to physics. Fortunately for the mental health of researchers everywhere, empirical research doesn't support Dirac's idea. One extensive study of Nobel Prize winners and similar innovators estimated that in today's world, a 50 year old has about 2.5 times more "innovation potential" than a 30 year old and that a 30 year old and a 60 year old have about the same potential to make a world-changing breakthrough. In the rest of the chapter, we look at several modern researchers who have achieved remarkable success and how long it took them to reach this state.

Fiona Wood and Spray-on Skin

Each year the Australian government designates a single Australian as the "Australian of the Year". In 2005, just a few years after Stephen Bradbury's stunning Olympic success, the Australian of the Year award went to a biomedical researcher and clinician, Dr. Fiona Wood.[1] As if that wasn't impressive enough, Dr. Wood was also voted the "most trusted Australian" every year from 2005 to 2010 in a *Reader's Digest* poll. Despite these accolades, Dr. Wood is the kind of doctor you definitely hope not to see as a patient – she is a plastic surgeon who specializes in treating patients with severe burns.

In her medical work, Dr. Wood was exposed daily to the pain experienced by burn victims and the laborious nature of treating burns with skin grafts. In the early 1990s, she started developing methods for rapidly applying layers of cells that could promote burn healing. Within a few years, the first tests of what became known as "spray-on skin" took place. In 2002 a terrorist bombing in Bali left many people horribly burned. Dr. Wood vaulted into the consciousness of the Australian public by successfully treating many victims of this bombing using a combination of spray-on skin and more conventional therapies.

Looking back on technical advances can easily convey a retrospective sense of inevitability. It is worth learning more about Dr. Wood's background to appreciate that her success probably seemed less than inevitable as she was achieving it. Fiona Wood was born in Yorkshire in 1958 in a home where neither of her parents had finished high school. Her parents encouraged her to pursue her education, a path that led her to medical school and training as a surgeon. In 1987 Dr. Wood, her husband (also a surgeon) and their two young children moved to Perth in Western Australia. During the years from which spray-on skin emerged, Dr. Wood completed her training as a plastic surgeon, maintained an active clinical practice and also had four more children. Next time you feel like you are a little short on time, pause and think about what life must have been like in the Wood family! Looking back over her career, Dr. Wood has given credit to advice from her poorly educated Yorkshire father, who said "The harder you work the luckier you get". She has also noted that "Things don't always go how you'd expect all the time. You have to pick yourself up and keep going. That's part of life".

---

[1] Dr. Wood is not the only scientist to win this award. The first Australian of the Year, in 1960, was Nobel Prize winner Macfarlane Burnet. Four other scientists have won the award since.

How long did it take for Fiona Wood to achieve success? By the time she moved to Australia in 1987, she had already completed multiple years of medical training. Twelve years later, in 1999, Dr. Wood cofounded a company to bring spray-on skin into widespread medical use. In 2017, 30 years after moving her young family to Perth, the company, now called Avita Medical, received FDA approval for its first therapy, a product called ReCell. Although no one would accuse Dr. Wood of being an overnight success, it seems likely that spray-on skin will indeed reduce the suffering of many burn victims in the future.

Lineages of Successful Scientific Papers

I am fortunate in my position at Georgia Tech to be surrounded by many incredibly creative and successful researchers. To delve into how long it can take ideas to incubate, I dug into the history of several of my colleagues' most notable papers. To warm up to this task, I started a paper from my own research group, not because my paper is especially insightful compared to the work of my colleagues but because I had lived the history myself. In 2002, then PhD student Anastasios Skoulidas, undergraduate student David Ackerman, my collaborator and friend Karl Johnson and myself published a paper in *Physical Review Letters* that described how fast molecules could move (or more precisely, diffuse) inside the tiny channels of carbon nanotubes. We showed that molecules could diffuse thousands of times faster than in any similar material, an outcome that was so surprising that we delayed writing the paper for quite some time while we checked and rechecked our results. This paper has now been cited in more than 500 other papers, a traditional, albeit imperfect, measure of a scientific paper's impact. By this measure, this paper is one of the biggest successes to come out of my research group.

For our purposes here I am less interested in what came after our paper than what preceded it. To characterize the work that preceded our 2002 paper, I looked at the chain of citations that could be constructed from the references in the paper. Every scientific paper uses references to highlight crucial preceding work, so a chain of references from paper to paper to paper can be constructed leading many decades into the past. I was interested in reconstructing the steps that were vital in preparing my collaborators and me to write our paper, so I focused solely on references to my own earlier work. There are three references of this kind in the 2002 paper, to papers I wrote with my collaborators in 2002, 2000 and 1999. The 2000 paper from this set was based on even earlier papers I had written, including references to a paper from 1999 and two papers from 1997. One of those 1997 papers was directly based on ideas from the very first paper I wrote as a PhD student in 1994. No one would classify some of those earlier papers as being enormous successes. My 1994 paper, for example, has been cited a grand total of 23 times in the "27 years" since it was published. But they all represent work that was vital to me developing the skills and ideas that made the 2002 *Physical Review Letters* paper possible. It is no exaggeration to say that this highly successful paper took more than eight years and a series of papers that attracted much less attention to come to life.

Whatever success my paper has seen certainly didn't occur overnight. But perhaps my career has just been on some kind of slow-track to success while my colleagues' careers were soaring like rockets. To test this unsettling hypothesis, I did the same analysis for several very successful friends in my academic department whose work I knew well. I began with Prof. Chris Jones, whose spectacular research in chemical engineering and chemistry has won him multiple prestigious awards. In 2008, Chris and his group published a paper in the *Journal of the American Chemical Society* describing how to make a new class of relatively cheap materials that are highly effective at capturing carbon dioxide, the primary greenhouse gas responsible for global warming. This paper has already been cited more than 570 times. More impressively to me, the paper has also led directly to development of real-world technologies. As of this writing, a startup company (the ambitiously named Global Thermostat) and ExxonMobil are working together to build a demonstration facility that will capture thousands of tons of carbon dioxide per year straight from the atmosphere using materials that directly stemmed from Chris' work.

How long did it take to do the work in that seminal 2008 paper? Prof. Jones and the first author of the paper (Jason Hicks, now a successful professor at the University of Notre Dame, who was a PhD student at the time) published two preceding papers in 2006. One of these relied directly on two papers by Prof. Jones and another student, Michael McKittrick, written in 2003 and 2004 and also on a paper Prof. Jones published while he was a PhD student in 1999. The latter paper, in turn, referred to another paper from Prof. Jones' work as a graduate student in 1998. This chain of references shows that my friend's masterful 2008 paper didn't come from an inspiration that was a bolt from the blue; instead, it was an outcome from at least a decade of sustained effort.

For a second example of a highly successful paper, I turned to work by my colleague Prof. AJ Medford. Most researchers go through their entire career without publishing a paper in *Science* or *Nature*. Prof. Medford achieved this milestone while he was still a PhD student in 2014 with an elegant paper in *Science* titled "Assessing the reliability of calculated catalytic ammonia synthesis rates". Because this paper was written with AJ's PhD advisor, Prof. Jens Nørskov, I tracked back through this paper's citation tree to papers by Prof. Nørskov. The 2014 *Science* paper builds directly on another *Science* paper from Prof. Nørskov and his group in 2005. This 2005 paper became extremely well known in the field; it was one of several key technical examples I highlighted in a technical textbook I coauthored several years later. More important for this discussion, this 2005 paper is part of a trail of citations to other work by the Nørskov group leading back to at least 1995. While my colleague's 2014 *Science* paper is without doubt a technical masterpiece, it is not an example of overnight success. Instead, it is one outcome (among many) from almost two decades of progress by a single research group.

I found the long lead times in my colleagues' work encouraging. Perhaps my own research success wasn't so slow after all – in fact, my years of work was starting to look less sluggardly. I decided to test this idea one more time by looking at the work of one more creative colleague, Prof. Hang Lu. Hang is part of a worldwide research community using a tiny worm called *Caenorhabditis elegans* (*C. elegans* to its friends) to probe a dizzying array of questions in bioscience. Researchers grow these tiny creatures by the thousands because their structure has been laboriously mapped in exquisite detail. Each worm has 302 neurons in its brain, for example, and every adult male has 1031 cells. They are amazingly resilient little creatures. They have been shown to readily survive exposure to accelerations of 400,000 g for an hour in an ultracentrifuge. *C. elegans* specimens reportedly survived the disintegration of the Space Shuttle Columbia in February 2003. *C. elegans* was the first multicellular organism whose genome was sequenced. Sydney Brenner, H. Robert Horvitz and John Sulston shared the 2002 Nobel Prize in Physiology for their work on organ development and cell death in *C. elegans*.

A biological experiment with *C. elegans* typically has a simple structure – a small change in the worm's genetics or environment is made and the effect on the creature's behavior is observed. This oversimplified description doesn't do justice to the hard work involved in doing these experiments. *C. elegans* can only be carefully observed through a microscope, so the observation part of the experiment involves someone, almost always a PhD student, staring intently into a microscope. A dedicated researcher willing to squint for hours can typically eyeball a couple of dozen worms on a good day. Because statistically significant conclusions demand large sample sizes and control experiments, embarking on a career studying *C. elegans* is a recipe for many, many days hunched over a microscope.

Prof. Lu decided to speed up *C. elegans* research by building an instrument to automatically measure and sort hundreds or thousands of worms without human intervention. In 2018, she published a paper doing just this in *Nature Methods* titled "Automated on-chip rapid microscopy, phenotyping and sorting of *C. elegans*". A key idea in Prof. Lu's work was to use microfluidics, a technology that adapted ideas from fabricating microchips to make devices that pump miniscule quantities of fluids like water through complex channels in highly controlled ways. The devices described in the paper sent hundreds of worms squirting through a miniaturized water park, directing them into different chutes based on their genetic properties. This approach sped up the rate of looking at *C. elegans* by many times, opening opportunities for experiments that earlier generations of researchers couldn't even dream about.

Once again I tracked back through the references in my colleague's breakthrough paper. To my surprise, it didn't contain a single reference to her earlier work. Had I finally found a genius who didn't need years of intellectual incubation to make a field-changing advance? I went to talk to Prof. Lu to find out. She explained that although there were no direct references in the paper, her work could not have been successful without years of background work, including five papers she wrote about microfluidics as a PhD student at MIT and another two papers on worm biology she wrote as a postdoc at Princeton University. This background meant that when Prof. Lu began her faculty position at Georgia Tech, she was probably the only person in the world who had done serious work in these two seemingly unrelated fields. These years of training were followed by more than three years of intensive work with two PhD students (Kwanghun Chung, a chemical engineer, and Matthew Crane, an electrical engineer), a period in which many worms gave their lives for science. The hard work wasn't finished once their microfluidic worm sorter worked. Their manuscript was rejected outright by two journals and then languished in review for five months before finally being accepted by *Nature Methods*. Just like the hugely successful papers by my colleagues Chris Jones and AJ Medford, the road to publishing a potentially career-defining paper took Hang Lu over a decade.

A tool like the microfluidic worm sorter can revolutionize a research area. But not everyone is ready for a revolution when it comes along. The worm sorter was initially received by polite indifference by the *C. elegans* community. Prof. Lu soon realized that biologists care about solving biology problems, not about new technology. This prompted her to collaborate extensively with a biology group to show how the worm sorter could help answer a question that couldn't be studied with traditional staring through microscopes. Three years after her original paper, the fruits of this work were unveiled in a 2011 paper in *Nature Methods*. The legacy of these years of work has been a steady adoption of worm sorting by *C. elegans* researchers worldwide, a powerful example of scientific impact.

There is a curious postscript to Prof. Lu's story. A few years into her work she received an email reminding her that it is not only biologists who are interested in worms. It read "We are a wholesale bait company in Ontario, Canada. We are looking at having a machine…to separate specific worms by size". Unfortunately, her correspondent didn't understand that the worms that are offered to hungry Canadian fish by anglers are hundreds of times larger than *C. elegans*. As far as I can tell, the bait suppliers of Ontario are still sorting their worms the old fashioned way, which presumably involves a lot of dirt and wet hands.

It might be tempting to characterize each of the scientific successes I have described here as a single paper by a professor and one or more graduate students appearing in a high-profile scientific journal. Said this way, it can look like these research groups have achieved overnight success. Digging a little deeper, however, showed that in every case, these papers were the result of sustained work that involved multiple researchers and that the typical time scale for this work was a decade or more.

The $10 Billion Blood Drop

My colleague Hang Lu wasn't the only person in the early 2000s trying to change how science was done by pumping tiny volumes of liquid around microfluidic devices. In 2003, a chemical engineering undergraduate dropped out of her studies at Stanford University to lead a company she had formed to improve diagnostic blood tests. Elizabeth Holmes' vision for her company, Theranos, was beguiling – they would replace the pain and cost of medical tests based on traditional blood draws with equivalent tests that would use only a single drop of blood from a pricked finger. Theranos developed an instrument called Edison that used microfluidics to move tiny amounts of blood in an automated fashion. As someone who feels queasy just seeing a picture of a hypodermic needle, I have to say that this new kind of medical test couldn't arrive too soon.

Although it didn't quite happen overnight, there can be no doubt that Elizabeth Holmes achieved amazing success. By 2013, Theranos had raised hundreds of millions of dollars in funding and was valued at over $10 billion. Holmes consciously modeled her wardrobe after uber-entrepreneur Steve Jobs and Theranos had all the trappings of an Apple-like trajectory. The company built a stunning headquarters in Silicon Valley, complete with a wall-sized display of wisdom from a guru of nerd culture, Yoda ("Do or do not. There is no try".). The company's advisory board included not one but two former US Secretaries of State, George Shultz and Henry Kissinger, a former US Secretary of Defense, William Perry, and a former US Senator, Sam Nunn. Holmes gave inspirational talks at TED conferences and other events and appeared on the cover of *Fortune* magazine (tag line: "This CEO Is Out For Blood") and *Inc.* magazine, where she was billed as "The Next Steve Jobs". Theranos closed major deals with the pharmacy chain Walgreens and the supermarket chain Safeway that would place Theranos diagnostic centers in easy reach of millions of consumers. Since annual sales in the USA alone for the diagnostic lab industry are more than $70 billion, the future of Theranos looked bright indeed.

If you don't already know the story of Theranos, you might be wondering why you can't get complex blood tests done routinely at your local pharmacy or supermarket. The reason is that despite all the money and magazine covers and TED talks, Theranos' technology never actually worked. In fact, it wasn't even close to working. At one level, this is not too surprising; all radically new technologies fail for some time before they succeed. The problems with Theranos, however, were vastly compounded because Holmes and other company leaders refused to publically acknowledge that any problems existed.

The unravelling of Theranos began in October 2015 when John Carreyrou, a reporter for the *Wall Street Journal*, published a series of articles questioning the truthfulness of Theranos' claims. Using information from a source within the company, Carreyrou revealed that medical tests performed with Theranos devices were highly unreliable and that tests purporting to drive their business agreements were being performed with traditional blood testing machines. At first, Theranos strongly denied these claims, but *The Wall Street Journal* articles led to an overlapping set of investigations by regulatory authorities, commercial partners and investors. In just nine months, Elizabeth Holmes' net worth collapsed from more than $4 billion to essentially zero. By July 2016, Theranos lost its license to operate a medical facility. After many months of legal wrangling and recriminations, Theranos closed permanently in September 2018. By this time hundreds of employees had lost their jobs (including at least one that committed suicide), multiple individual investors had each lost more than $100 million, and most seriously a large numbers of consumers were given medical test results that were unreliable. The human cost of these events is described in far greater depth in John Carreyrou's gripping book *Bad Blood.*

The bad news for Elizabeth Holmes didn't end with her fortune vanishing and her public credibility collapsing. In early 2018, she and the president of Theranos, Ramesh Balwani, were each charged with fraud by the US Securities and Exchange Commission. Among other allegations, the SEC accused the pair of reporting Theranos as having annual revenues of $100 million when the true number was only $100,000. In July 2018, the pair were also indicted by federal prosecutors on wire fraud and conspiracy charges. As of this writing, they are scheduled to go on trial in 2021.

The full motivations that led Elizabeth Holmes from being a creative young entrepreneur to standing trial for being a systematic liar and fraud may never been entirely known. I suspect that this calamity didn't arrive in one terrible decision. Instead, an initial pitch to investors about a speculative technology led to overconfident predictions, which led to fudging a little data to keep someone happy, which in turn demanded bigger and bigger lies.

Why does the story of Theranos belong in a chapter about becoming an overnight success? If you have ambitious aims in your creative pursuits, your colleagues will typically admire and celebrate you, even if the full scope of your ambitions aren't met. Ambition, however is just one of multiple elements that are required for any long-term success. The rise and fall of Theranos illustrates in stark terms the imperative of integrity in your life and career. Dishonesty of any kind will be toxic to your reputation and career. Resist the temptation to taken even the smallest step down any path that requires you to compromise your integrity.

Elizabeth Holmes cast a vision that inspired many people. Hundreds of millions of dollars were spent trying to make that vision a reality. Ironically, the same magazine cover that proclaimed Holmes "The Next Steve Jobs" also had another cover story titled "How To Build A Company To Last 100 Years". The deceit that lay of the heart of Theranos meant that the company fell tragically short of this goal.

Kuen Charles Cao and the Invention of Optical Fibers

It would be a shame to end this chapter with the discouragement and duplicity of the Theranos saga, so we will look at one more instance of truly world-changing research. The ubiquitous role of electronic data in our lives would not be possible without optical fibers. These glass fibers are the "pipes" that speed huge volumes of data over vast distances at the speed of light. When I first moved from Australia to the USA in the early 1990s, trans-Pacific telephone calls often featured halting conversations because of the time delay associated with connections via telecommunications satellites. Today, undersea optical fiber cables have reduced this problem to the shortest limit allowed by the laws of physics.

In the early 1960s, the only option for sending data via cable was to use copper wires, a technology that had not changed greatly since the invention of the telegraph. Although the volumes of data used then look meager now, electronics engineers were already anticipating the need to move beyond copper wires. In 1960, T. H. Maiman and coworkers at the Hughes Research Laboratory in Malibu, California, demonstrated the first ever laser. Almost immediately, several groups realized that if a laser could be transmitted efficiently over long distances with minimal losses this would provide an ideal way to deliver data.

In 1963, Kuen Charles Kao was working at Standard Telecommunications Laboratories in England. Kao's group had started to send laser light through glass rods, early precursors of optical fibers. Their experiments showed that the laser barely propagated at all. Rather than being discouraged, Kao made a bold hypothesis that the signal attenuation was due to impurities in the glass. If he was correct, and it seems that there was very limited evidence that he was, this idea implied that progress could be made by learning how to make purer glass.

Kao's hypothesis drove an intensive period experimenting with techniques to make purer glass. Remarkably, it was only during this period that Kao completed his PhD, working as an external student at University College London. After three years, Kao had done experiments with rods that showed attenuation of 1000 dB/km. To compete with copper wire, this attenuation needed to be reduced to less than 20 dB/km. The scope of this challenge is undersold by using these unfamiliar units, which have little intuitive meaning to most of us. Because decibels (dB) are a logarithmic scale, Kao and his team needed to improve their performance by a factor of 10^{98} to achieve their goal! Remarkably, this situation didn't daunt them or, apparently, the people approving their funding. With three years more work, Kao reduced losses to the equivalent of just 4 dB/km, albeit only in a glass rod 1 m long. This demonstration showed the world that optical fibers had the potential to replace metal cables for electronic communication.

Just one year later, in 1970, the Corning company produced an optical fiber more than 1 km long with losses of less than 20 dB/km. Ten years after the invention of the laser and seven years after Kuen Charles Kao's insightful hypothesis, optical fiber communication was starting to seem possible. Thirty-nine years later, in 2009, Kao shared the Nobel Prize in Physics for his work. Of all the scientific accomplishments we have considered in this chapter, the creation of optical fibers is probably the most speedy. In only seven or eight years, Kuen Charles Kao had led an effort that went from what must have seemed like a wildly speculative idea to a working prototype.

The story of optical fibers doesn't end in 1970. Even though the potential of the technology was clear, making high-quality optical fibers in large quantities was far too expensive for widespread deployment. To compete with copper wires, it had to be possible to make many thousands of kilometers of fibers at costs similar to copper wire. In 1983, a chemical engineer working at Corning, Thomas Mensah, made dramatic advances in the speed at which optical fibers could be manufactured. With this breakthrough, optical fibers were cost competitive with copper cables for the first time. Because of his contributions to optical fiber manufacturing, among other accomplishments, Thomas Mensah was elected to the US National Academy of Engineering. He was one of the first African-Americans to receive this accolade.

Are You Ready to "Do a Bradbury"?

This chapter opened with the story of Stephen Bradbury's remarkable Olympic gold medal in 2002. Ironically, the man whose exploits were the origin of the phrase "to do a Bradbury" didn't come by his success easily or overnight. Bradbury first became a competitive skater in his early teens. It would be hard to overstate how far outside the mainstream this activity was. Today, there are fewer than 25 ice rinks in Australia and there were considerably fewer in the 1980s as Stephen Bradbury was spending his afternoons, evenings and weekends on the ice. Despite the fringe nature of ice skating in Australia, the speed skating World Championships took place in Sydney in 1991. Eighteen-year-old Bradbury was a member of the Australian men's 5000 meter relay team, and his team rode a wave of hometown enthusiasm to win a bronze medal. This was the first medal of any kind by Australia in the world championship of any winter sport. Although their achievement gave Australian skaters great confidence heading into the 1992 Winter Olympics, Bradbury wasn't selected to race and watched the Olympics from the sidelines. He continued his training and just two years later was again a member of the 5000 meter relay team in the 1994 Olympics, the first time the Winter Olympics were held out of sync with the quadrennial Summer Olympics. Bradbury and his team again made antipodean winter sporting history by earning a bronze medal, the first medal of any kind by Australians in the Winter Olympics.

At the age of 21, Bradbury was already Australia's most decorated winter athlete. Later that year, Bradbury was skating in a World Cup event when he fell in a collision with another skater. One of the other athlete's skates severed an artery in Bradbury's leg, and it is estimated that he lost 4 liters of blood before being stabilized. His injury required more than 100 stitches and immobilized him in hospital for three weeks. He continued training and was able to qualify once again for the Olympics four years later in 1998, where he competed in three events but did not earn any medals. Two more years of training followed. In 2000, Bradbury fell in training and broke his neck. One doctor advised him "you will never skate again", and Bradbury spent six weeks in a halo brace before he was allowed even limited movement of his head.

When thinking about the six years and two life-threatening accidents that had passed since his Olympic medal, no one would have second guessed Bradbury if he decided it was time to hang up his skates. Amazingly, he continued with his training. By 2002, he was racing for Australia in his third Olympics in the event that would soon bring him worldwide fame, the 1000 meter short track. Winning the short track event required racing at least three times on the same day, once in a heat, again in the semi-finals and the final. Bradbury knew that he didn't have the speed or stamina to beat his much younger rivals, even if he raced perfectly. In the semi-final, he adopted a strategy that must have felt like desperation – he deliberately skated behind the race leaders in the hopes that one or more of them would crash. His strategy paid off when three front runners slid off the track, clearing the way for the Australian to qualify for the final. He had "done a Bradbury" just to qualify for the Olympic final. The rest is history. By necessity, Bradbury used the same strategy in the final and presumably could hardly believe his eyes as once again all the skaters ahead of him toppled off the track and he cruised across the finish line to win a gold medal. When asked how he felt about what the world saw as overnight success, he said "I'll take it for the last decade of slog that I put in". His gold medal certainly involved an element of luck, but his persistence and strategy put him in a position where luck could tip the scales of success in his direction.

Stephen Bradbury's overnight success came after more than 15 years of dedicated training, mostly in complete obscurity. His thousands of training sessions must have been driven by a day-to-day ability to set goals and derive satisfaction from incremental milestones. World changing advances in research and creativity work in this way too. Are you ready to do a Bradbury? In the rest of this book, we explore some of the skills and habits that will help you succeed if this is path you have set for yourself.

Chapter Summary

- Dramatic breakthroughs in research rarely occur as "bolts from the blue" with no precedent or warning. Instead, exciting advances are usually preceded by years of work that, at the time it was performed, seemed incremental and uncertain.

- Lapses in integrity can have long-lasting and devastating professional consequences. Don't let ambition or perceived outside forces tempt you to act unethically in your work.

- To have the chance of "doing a Bradbury" with your research, a commitment to steady hard work over long periods of time (years, not days or weeks) is needed.

Further Resources for Chapter 1

Stephen Bradbury's remarkable life story was described in detail in a long-form interview by Richard Fidler with the ABC Radio program *Conversations* in December 2017 that is available online: https://www.abc.net.au/radio/programs/conversations/conversations-steven-bradbury/9199060. This link also includes a way to view a video of Bradbury winning his 2002 Olympic gold medal. Dr. Fiona Wood was interviewed on the same program in 2012: https://www.abc.net.au/radio/programs/conversations/dr-fiona-wood-continues-her-remarkable-skin-regeneration-work/7757088

The study that explores the connection between age and accomplishment in various fields is Age and Great Invention, B. F. Jones, *The Review of Economics and Statistics*, 92 (2010) 1–14.

The four papers whose "lineage" was discussed in the chapter are Rapid Transport of Gases in Carbon Nanotubes, A. I. Skoulidas, D. M. Ackerman, J. K. Johnson and D. S. Sholl, *Physical Review Letters*, 89 (2002) 185901, Designing Adsorbents for CO_2 Capture from Flue Gas – Hyperbranched Aminosilicas Capable of Capturing CO_2

Reversibly, J. C. Hicks, J. H. Drese, D. J. Fauth, M. L. Gray, G.-G. Qi and C. W. Jones, *Journal of the American Chemical Society*, 130 (2008) 2902, Assessing the Reliability of Calculated Catalytic Ammonia Synthesis Rates, A. J. Medford, J. Wellendorff, A. Vojvodic, F. Studt, F. Abild-Pedersen, K. W. Jacobsen, T. Bligaard and J. K. Nørskov, *Science*, 345 (2014) 197 and Automated On-Chip Rapid Microscopy, Phenotyping and Sorting of *C. elegans*, K. Chung, M. M. Crane and H. Lu, *Nature Methods*, 5 (2008) 637.

A key figure in the downfall of the Theranos was the *Wall Street Journal* journalist John Carreyrou. His book *Bad Blood: Secrets and Lies in a Silicon Valley Startup* (Alfred A. Knopf, 2018) is a gripping account of Theranos' rise and fall. Unfortunately, Theranos is not the only company that has soared to enormous valuations only to collapse because of dishonesty by the company's leaders. An example from roughly a decade earlier is the history of Enron, which is detailed in *The Smartest Guys in the Room: The Amazing Rise and Scandalous Fall of Enron*, Bethany McLean and Peter Elkind (Portfolio Trade, 2003). This book was the basis of a 2005 documentary, *Enron: The Smartest Guys in the Room*.

2

Deep Work, Shallow Work and Frippery

As you begin to read this chapter, I invite you to participate in a short experiment: cut yourself off from electronic connections with the outside world for 30 minutes. If you are reading this from a physical book, turn off your phone and computer, find a comfortable chair and concentrate simply on reading for 30 minutes. If you are reading on an electronic device, switch it to airplane mode and don't use anything other than your reading app. Does this experiment sound difficult? Exhilarating? Old-fashioned? Perhaps you feel like Nicholas Carr, who wrote in his 2010 book, *The Shallows: What the Internet Is Doing to Our Brains*, "Over the last few years I've had an uncomfortable sense that someone, or something, has been tinkering with my brain….I'm not thinking the way I used to think. I feel it most strongly when I'm reading. I used to find it easy to immerse myself in a book or a lengthy article…That's rarely the case anymore….I get fidgety, lose the thread, begin looking for something else to do".

Osamu Shimomura shared the 2008 Nobel Prize in Chemistry for his role in the discovery of green fluorescent protein (GFP), a molecule that glows a bright green and is now used in thousands of laboratories around the world as a vital tool in biological experiments. This success must have been hard to predict in 1945 when the then 16-year-old Shimomura was temporarily blinded by the detonation of the world's second atomic bomb over Nagasaki, 25 km from his home. In 1960, Shimomura was a researcher at Princeton University and was given the task of understanding the bioluminescence of jellyfish that were found in the coastal waters of Washington State. This required collecting jellyfish, lots and lots of jellyfish, so that various chemicals could be extracted from them. Shimomura later described his daily routine during the summer season in Washington: collecting jellyfish from six to eight o'clock in the morning, cutting rings of tissue from the jellyfish until noon and then spending the afternoon extracting chemicals from the tissue samples. After a dinner break, he spent two more hours collecting jellyfish. That probably isn't what you think of when you imagine a relaxing day at the beach. In the first summer he spent in Washington, Shimomura and the others working with him collected around 10,000 jellyfish.

Shimomura's routine of long summer days collecting and processing jellyfish continued for years, and these efforts were matched by long months in the lab in Princeton performing chemical purification experiments. By 1966, six years after his first trip to Washington, he had isolated the luminescent agent, now called aequorin, and felt that it should be possible to determine the molecule's structure. This experiment would require 100–200 mg (less than 1/10th of a teaspoon) of purified aequorin. Obtaining this sample, however, would require processing roughly 50,000 jellyfish. This realization motivated the development of various tools such as mechanical cutting machines to catch and process thousands of jellyfish a day and improvements in lab-based purification methods. These development took roughly 20 years to come to fruition, during which Shimomura collected more than a million jellyfish.

It is worthwhile pausing to compare Osamu Shimomura's years of patient work to Nicholas Carr's description of having trouble maintaining the focus necessary to read a book. Which one feels more like your day-to-day efforts to focus on your work? Although Shimomura's decades-long quest to understand a bioluminescent molecule is an extreme example, it further highlights the observation from Chapter 1 that making substantial advances in any creative endeavor requires sustained attention and focus. In this chapter, we will discuss some of the patterns of thinking and day-to-day habits that can help you work in this way.

The title and many of the ideas in this chapter are adapted from Cal Newport's excellent book *Deep Work: Rules for Focused Success in a Distracted World*. Newport is an academic computer scientist at Georgetown University who studies abstract theories of dynamic networks (sample journal article title: "Contention resolution on a fading channel"). Driven by his personal interest in achieving academic success while maintaining something resembling a normal life and family, he has thought carefully about work habits and collected these ideas in a series of widely read books. What is deep work? Newport defines it as

> professional activities done in a distraction-free state
> that push cognitive abilities to their limit.

All three parts of this definition are important. The focus on "professional activities" reminds us that this kind of work isn't done just for its own sake but is a core element of being successful in careers based on creative pursuits. The necessity for a "distraction-free state" hints at how easily distractions can destroy a complex train of thought. Finally, to accomplish something truly original and meaningful in any field of research is hard, so it can only be done if you are able to "push cognitive abilities to their limit".

The characteristics of deep work contrast with shallow work, which Newport defines as

> undemanding logistical-style tasks, often performed while distracted.

Shallow work isn't bad or unimportant. Ordering routine supplies for your lab so experiments can continue is shallow work. Recording the homework grades for students in the class you are teaching is shallow work. So is booking travel arrangements for the conference where you are giving a talk or completing the online ethics training your institution mandates annually for all employees. Failing to complete any of these tasks in a timely way can have intensely negative consequences, so they can't just be ignored. They don't, however, push your cognitive abilities to their limit or require complete and distraction-free concentration to be successfully completed.

Another characteristic of shallow work tasks is that spending more than the minimum amount of time they require is unlikely to yield much positive benefit. An exhaustive comparison of plane ticket options for your conference travel or a deep dive into the cancellation policies of the multiple hotels near the conference center isn't going to improve the conference talk you are preparing. Creating and using a complex multi-tiered grading structure with pens of three different colors is almost certainly not going to help the students in your class get more out of their homework assignments. Achieving a state in which you perform no shallow work is not a sensible goal, but it is sensible to find ways to minimize the time you spend to adequately complete the shallow work that must get done.

The definitions of deep work and shallow work make it clear that both kinds of work need to be part of a creative, research-based career. There is a third category of time use, however, that we all experience: frippery. This is an old-fashioned word that carries the ideas of something that is trifling, frivolous or empty. Playing a mindless game on your phone is frippery, as is following an acrimonious dispute among celebrities on your favorite social media platform. Completing a daily crossword may feel somehow more virtuous than Candy Crush or Instagram, but this too can be frippery. If an activity has no bearing at all on your professional work and you would feel vaguely embarrassed if someone measured the time you spent on it there is a good chance that this activity can be described as frippery.

By labelling some of the distractions that divert us as frippery I am not trying to say that all diversions are bad or that you should aim to become an automaton who works constantly to the exclusion of everything else. Finding ways to relax and recharge our mental and physical resources is important for everyone. It is useful, however, to explicitly identify these activities so you can allot time to them consciously instead of feeling that an entire afternoon or evening has disappeared without quite being able to tell what you did with your time. Just as importantly, we must realize that the habits we develop when we are not working also deeply influence us when we are working. This is the idea Nicholas Carr pointed to in the subtitle of his book, *What the Internet Is Doing to Our Brains.* Carr argued that constant exposure to a connected environment where clicking through to another link is far easier than thoughtful reading has broadly eroded our capacity for creative reflection and connection. To say this another way, repeated exposure to the stimulus overload of many Internet-enabled activities makes it fundamentally harder for us to maintain the focus that is a prerequisite for any kind of deep work.

The idea that spending time engaging in distraction-heavy activities subtracts from our long-term abilities to think clearly is not a new one. In 1985, long before cell phones and social media were widespread, Neil Postman made many of the same points as Nicholas Carr in his book *Amusing Ourselves to Death*. The target of Postman's displeasure seems almost quaint today; his book was about the negative impacts of television on society. Postman contrasted public discourse before television ("generally coherent, serious, and rational") and in the age of television ("shriveled and absurd"). He had prescient views on the dangers of conflating education and entertainment. His tongue-in-cheek commandments for what might be called edutainment included "Thou shalt have no prerequisites", "Thou shalt induce no perplexity" and "Thou shalt avoid exposition like the ten plagues visited upon Egypt".

There is one element missing from the definition of deep work given above. To reap the benefits of deep work, this kind of work must be performed persistently over long periods of time. This was one of the major themes of Chapter 1: inspired creative work requires steady effort over months and years of concentration. A striking historical example from my own institution, Georgia Tech, illustrates some risks of trying to sidestep this aspect of scientific progress. On March 23, 1989, Martin Fleischmann and Stanley Pons from the University of Utah held a press conference to announce an incredible breakthrough; they had observed nuclear fusion in a table top device at room temperature. Generating energy from fusion, the energy source that powers the sun, had been a target of the physics community for decades. Traditional work on fusion aimed to mimic the incredible temperatures found within the sun to overcome the enormous repulsion that exists between the components of atomic nuclei. The stunning process outlined by Pons and Fleischmann involving a simple electrochemical cell in a regular chemistry laboratory rapidly became known as cold fusion. In a hint of the problems to come, the University of Utah work had not been peer reviewed or published at the time of the March 23 news conference, although the university had filed paperwork to protect its patent rights.

In an astonishing historical coincidence, the day after Pons and Fleischmann's news conference saw another dramatic energy-related event. On March 24, the oil tanker *Exxon Valdez* ran aground in Alaska, spilling more than ten million gallons of oil. The contrast between the pollution and risk associated with fossil fuels and the promise of seemingly limitless clean fusion energy was irresistible. Labs around the globe began a frenzy of work to repeat and improve upon the original cold fusion experiments, and Georgia Tech was no exception. On March 29, just six days after the University of Utah news conference, Bill Mahaffey submitted a $25,000 research proposal to the Georgia Tech Research Institute (GTRI). Within days funds had been made available, and Mahaffey, Gary Beebe, Darrell Acree, Rick Steenblik and Bill Livesay were working around the clock gathering supplies and building equipment. On Saturday April 8, the Georgia Tech cold fusion experiment started its first runs. There can be little doubt that the experimenters were ignoring distractions and pushing their cognitive abilities to the limit; that is, they were engaged in deep work.

After a weekend of intense experiments, the GTRI team held their own press conference on Monday April 10, announcing that they too had observed cold fusion. In particular, they had detected neutrons being emitted from their electrochemical cell, a key signature of fusion. Their results attracted national media attention. In a front page story the next day from the *Atlanta Journal*, Bill Livesay was quoted as saying "There's no question it's fusion. I still don't believe it, even though I see it" and Bill Mahaffey added "It happened so soon, we thought it was an equipment malfunction". The same article included admiring quotes from scientists at MIT and the Los Alamos National Laboratory together with a comment directly from Stanley Pons, who said "This is just incredible, I can't remember when I've had a better Monday".

Unfortunately, the Georgia Tech team's elation was short-lived. Continued tests convinced them that their neutron detector was only registering a signal because it had heated up, not because it was actually counting neutrons. On Thursday, only three days after their first news conference, Georgia Tech made the difficult but admirable decision to call a second news conference. The following day they were once again the subject of stories in the *Atlanta Journal* (a front page piece headlined "Tech Scientists Retract Fusion Claim") and *The New York Times* ("Georgia Tech Team Reports Flaw in Critical Experiment in Fusion"). This is presumably not the kind of publicity you hope your own research to generate!

In less than a year the scientific community came to a consensus that Pons and Fleischmann's cold fusion was not fusion and did not generate energy. Not surprisingly, researchers at Georgia Tech were not the only ones to succumb to the allure of potential short-term fame and fortune via cold fusion. An article from the *Atlanta Journal* just a few days after the Georgia Tech retraction interviewed a graduate student from Duke University who had been busy building cold fusion cells "and missed his girlfriend's birthday party in the process". The student, James Langenbrunner, was said to be reluctant to return to his "pre-cold fusion work in nuclear physics", quoting him as saying "You don't know what a drag it is to use that accelerator. You can work for days without getting any usable data at all". Ironically, after completing his PhD, Langenbrunner went on to have a successful career at the Los Alamos National Laboratory where he published work, among other things, on inertial confinement (i.e., "hot") fusion. Apparently, he realized the long-term value of deep work and that working for days was typically a necessary part of generating valuable scientific data.

Why Is Deep Work Worthwhile?

The long-term value of deep work relative to shallow work or frippery is easy to agree to. You probably don't know many people who will argue that spending more time looking at their cell phone or surfing the Internet will make their research more productive. So why does adopting a lifestyle focused on deep work seem so difficult? Before we explore habits that can improve your ability to accomplish deep work, it is helpful to more carefully consider why (and whether) deep work is worthwhile.

The most pragmatic reason to aspire to deep work is that the world is a competitive place. People with special skills are far more likely to be rewarded than people who are just average. If Neil Postman, Nicholas Carr and Cal Newport are right that modern media are steadily diminishing our collective ability to concentrate and think, then having that ability will become even more valuable over time. As Cal Newport put it, "Our culture's shift towards the shallow is exposing a massive opportunity for the few who prioritize depth". In today's hyperconnected world, access to information is rarely the limiting factor in making research breakthroughs. Instead, the ideas that lead to dramatic shifts in the progress of a research field typically involve looking at existing information in fresh ways or asking more insightful questions. The skills needed to achieve these kinds of successes are in short supply, even in the world of creative research. Developing these skills will make your work stand out from the crowd and give you a competitive advantage in a competitive world.

One example of deep work leading to insights that had eluded others comes from the work of Alfred Gilman and Louis Goodman. In the 1930s, Gilman and Goodman were new Assistant Professors at Yale Medical School who were assigned to teach pharmacology, the study of how drugs of all kinds affect patients. In an experience common to professors across the ages, they weren't satisfied with the books that were available to teach their course. They decided that what was needed wasn't just another book, but a book that took a fundamentally different approach to all earlier work in the field. Specifically, they wanted to carefully examine the evidence for efficacy of each therapeutic drug they discussed. This evidence-based approach may sound obvious in retrospect, but it was a new idea at the time and it became the basis for a monumental research and writing project. Much to the consternation of their publishers, Gilman and Goodman eventually delivered a manuscript with more than a million words. For comparison, a typical novel contains around 80,000 words. Gilman and Goodman had completed in just a few years what for many people would be the research and writing output of a lifetime. The book *The Pharmacological Basis for Therapeutics* appeared in 1941, and it rapidly redefined how the field was taught and practiced. The book's publisher initially printed 3000 copies and promised the authors a case of scotch if those copies sold within four years.[1] The authors collected their scotch after only six weeks of sales and the book went on to sell over 86,000 copies. Gilman's son, Alfred Goodman Gilman, was born in the same year and was given his

---

[1] This kind of incentive appears to have disappeared in present day scientific publishing.

middle name to honor his father's coauthor.[2] Almost 80 years later, the book is in its 12th edition and remains a touchstone in its field.

Their textbook was not Gilman and Goodman's only lasting legacy in medical science. After his appointment at Yale, Gilman worked for the US Government seeking to understand the biological effects of mustard gas, a chemical warfare agent from World War I. He collaborated with Goodman to measure the toxicity of mustard gas on white blood cells in mice. The pair had the audacious idea that this toxicity could be used for good in treating cancer. Within a few years, they had completed a clinical trial and introduced what is now viewed as the first chemotherapy drug. It seems unlikely that Gilman and Goodman would have made the creative leaps needed for this breakthrough without the deep work that immersed them in the history and science of pharmacology during the writing of their magnum opus.

It would be a mistake to assume that everything Gilman and Goodman touched turned to medical gold. By the mid-1940s, Louis Goodman had moved to the University of Utah, where he was interested in developing anesthetics. He decided to test the anesthetic properties of curare, a paralytic agent used as a poison in South America. In a study that would horrify modern ethics review panels, Goodman persuaded a colleague to be injected with a small dose of curare, after which the incapacitated patient would communicate by blinking. Before the experiment, it was not known if the paralysis from curare was temporary or not, or whether a paralyzed individual could feel pain. After receiving the injection, Goodman's paralyzed colleague indicated he could still feel pain when he was poked with pins, removing any hope that curare could be used as a clinical anesthetic. The experiment went from bad to worse when the patient stopped breathing. Goodman kept his colleague alive by inflating his lungs using a rubber bag. Fortunately for all concerned, the curare paralysis was temporary and the whole incident became nothing more than an alarming story for future medical students.

The ability to perform deep work in a sustained way creates a competitive advantage in a competitive world. This situation exists because deep work is a rare skill. Working on a project with the habits of deep work requires a level of personal commitment and motivation that many people are unable (or unwilling) to muster. It is tempting to look back on the years of work by Osamu Shimomura purifying chemicals from jellyfish or by Alfred Gilman and Louis Goodman writing their million word textbook through the lens of their eventual success. But at the time the work was being done, the success of these projects was far from assured. Being unable to chart a path through long periods of hard work halts many creative projects before they are started.

---

[2] In an instance of the apple not falling far from the tree, the younger Gilman was also a highly successful researcher who later shared the Nobel Prize in Physiology or Medicine in 1994 for his discovery of G-proteins. A colleague joked that Gilman was "probably the only person named after a textbook".

An example of the rarity of deep work comes from the life and accomplishments of Marie Curie. Working at a time when science was almost exclusively a domain reserved for men, Marie Curie became the first woman to win a Nobel Prize and remains the only person to have won a Nobel Prize in two separate fields of science. Marie Curie hypothesized in the 1890s the existence of radioactive elements whose radiation was more intense than uranium. This idea motivated years of experiments by Marie Curie and her husband Pierre in a laboratory that was little more than a leaky shed. By 1902, this work had led the Curie's to purify 0.1 g of radium chloride from more than two tonnes of the mineral pitchblende.[3] Despite the prejudices that existed about the possibility of a woman being a scientific leader, it appears that Marie Curie was the creative force behind the success of the Curie's efforts. One account at the time described Marie as Pierre's "greatest discovery". In June 1903, the University of Paris granted Marie Curie her doctorate, a decision that the university administration probably was relieved by when in December of the same year she shared the Nobel Prize in Physics with her husband and Henri Becquerel.

After Pierre Curie's death in a traffic accident in 1906, Marie Curie continued her research and received the Nobel Prize in Chemistry in 1911 for her discovery of radium and polonium. During World War I, she developed the world's first mobile X-ray machines, which were used to treat more than a million soldiers. Marie Curie died from the effects of radiation poisoning, which were unknown at the time of her research, in 1934. The papers from her laboratory are considered so radioactive by modern standards that they are stored in lead-lined boxes.

This brief summary of Marie Curie's work could give the inadvertent impression that she made a "lucky" hypothesis and stumbled upon a scientific discovery. Such a description of her success would be completely incorrect. Her initial insights were derived from careful experiments with natural minerals and careful contemplation of the current state of knowledge about uranium. These ideas motivated carefully planned experiments that took years to complete in physically unpleasant surroundings. These characteristics exemplify the rarity of deep work.

Chopping up tens of thousands of jellyfish. Spending years in a leaky shed to separate a speck of a new compound from tons of rocks. Writing a million word textbook on the biological effects of drugs. Contemplating these examples, you could be forgiven for concluding that deep work is valuable and rare but it doesn't sound like a lot of fun. This brings us to the third and probably most important reason that deep work is worthwhile: deep work is satisfying. As I hope you have already experienced in your own work, there is enormous pleasure in creating something new or in skillfully completing complicated tasks. This kind of satisfaction was

---

[3] The scale of the initial material and final product in these experiments is amazingly similar to Osamu Shimomura's experiments with jellyfish more than 60 years later.

famously characterized as "flow" by a Hungarian-American psychologist with a tongue-twisting name, Mihály Csíkszentmihályi. Csíkszentmihályi has described flow as "being completely involved in an activity for its own sake. The ego falls away. Time flies....Your whole being is involved, and you're using your skills to the utmost". There is a strong overlap between this description and Cal Newport's definition of deep work: "professional activities done in a distraction-free state that push cognitive abilities to their limit". Flow can certainly be achieved in non-professional as well as professional activities, but one of the key privileges of any career focused on research or creative pursuits is that it rewards and encourages the ability to achieve flow.

A key aspect of flow is that someone experiencing it finds their activity intrinsically rewarding. If you want a fancy word for this to impress your friends, you can describe it as an autotelic experience. In this state, you are motivated to push your work higher and further simply because it feels good. The results of this kind of deep work might bring external recognition months or years later, but in the moment doing good work is its own reward. I view this as by far the most compelling reason to aim to do deep work rather than shallow work or frippery.

Mihály Csíkszentmihályi first developed his description of flow while studying painters who got so engrossed in their work they would forget to eat or drink. These subjects point to a sensible objection to this whole deep work thing: that sounds fine for a painter in a sunny studio somewhere, but in my real life who is going to get my kids to school, grade last week's exam in the class I am teaching, take my car to the repair shop and fill out the paperwork my funding agency is demanding from me? Everything on this hypothetical to-do list, and many similar items from your own life, is vital to the health and well-being of your family and your ability to continue holding a job. They also fit within the definition given above of shallow work and remind us that doing shallow work well is also worthwhile. It isn't sensible to imagine that you can fill all of your time with intensely satisfying deep work and reduce the fraction of time you spend on shallow work or frippery to zero. But the intense short-term satisfaction and long-term benefits you can reap make it is worth learning how to immerse yourself in deep work as a routine part of your working life. This leads us to our next topic: training yourself to do deep work.

Making Deep Work Work for You

Our lives are surrounded by potential distractions. Most of us carry a smart phone that can instantly connect us to a myriad of apps and sites that aren't just potentially distracting; they have been actively designed to capture and feed our short-term attention. Remember the experiment I invited you to be part of in the first paragraph of this chapter? If you made it this far without distraction, congratulations! If you didn't, make a list of the ways you have used a network or Internet connection since you started reading the chapter. How many of those things would improve your concentration if you were to use them while trying to do deep work?

In the rest of this chapter, we will explore three strategies that will help you be better at doing deep work. These are not feel-good "success in 15 minutes" strategies; they require steady commitment and practice. The skills of deep work can be learned; they are not intrinsic personal qualities that you either have or don't have. Athletes who perform at the highest levels, like the ice-skater Stephen Bradbury who we met in Chapter 1, hone their skills over years of repeated practice. Creative researchers who benefit from the ability to perform deep work are similar. As you think about how to use the strategies we discuss below in your own life, please don't aspire to an idealistic state of perfection where every day is a model of positive work habits and frippery-free living. Instead, try some specific changes in how you approach your work, making sure you stick with the changes long enough to test the outcomes. By trying these behavioral experiments, you will find ways to train yourself to reach higher and higher levels of performance over time.

Establish Schedules and Rhythms for Your Work

In the previous section, we discussed flow, the intensely enjoyable experience of being deeply immersed in a challenging task. A problem with the idea of flow is that it describes what the experience of being in the middle of deep work feels like but it doesn't say anything directly about how to actually get into that state. Said more prosaically, flow feels great once you are in it, but it is hard to start. Even if you have the self-motivation to start a distraction-free session of intense work, how do you find the time? None of the life and professional obligations we discussed above that fill your to-do list with important but shallow work are going to magically complete themselves, and many of them have real deadlines and consequences associated with them.

This lament covers two of the top reasons that almost everyone sees as barriers to doing deep work, namely, "I just don't have time" and "Even when I have time I can't get started". A key to pushing past these barriers is remarkably simple: schedule the times you use for deep work. This sounds thoroughly uncreative, but it is amazingly effective.

An example of this scheduled approach comes from the life of H. L. Mencken, a prominent American writer and social critic during the first half of the 20th century. Today Mencken is primarily remembered for his witty aphorisms. To give just two of many possible examples: "For every complex problem there is a solution that is clear, simple, and wrong" and "A man may be a fool and not know it, but not if he is married". Mencken wrote more than 30 books, including a multi-volume study of American use of English, and was a nationally syndicated newspaper columnist. He also received and wrote hundreds of thousands of letters during his life.

How did Mencken get all of this work done? He had an organized schedule of tasks that he tackled each day. In the morning, he focused on reading and answering his mail, while the afternoon was devoted to editorial duties. Notice that this filled his morning and afternoon with shallow work activities that Mencken certainly needed to attend to but didn't directly contribute to his core output as a writer. Mencken's habit was to complete a session of concentrated writing each evening before he stopped work after a long day.

Mencken's approach highlights several positive aspects of adopting a regular rhythm for deep work. First, and probably most critically, his writing sessions were a fixed habit in his daily schedule, freeing him from any internal debate about whether he "felt like" writing on a particular evening. Over a career that lasted year and then decades, it seems likely that Mencken's evening writing became a core part of his self-identity: it is evening, so it is time for me to sit at my desk and write.

Second, Mencken's writing involved a block of time long enough that he could make real progress but not so long that he became mentally exhausted and inefficient. Spending six chunks of 15 minutes spread throughout a day cannot be as effective for deep work as a single focused session of 90 minutes. Becoming fully engaged in the concentration needed for deep work is hard, and it takes time to "warm up" before the best results can occur. Deep work, as its definition says, pushes cognitive abilities to their limit. Even people who have developed high levels of mental stamina can't maintain this kind of focus for more than a few hours.

Mencken's daily schedule strongly suggests that he had control over how he spent his time. This brings us to the third positive aspect of Mencken's approach: he organized his day so that his deep work was scheduled when he was best equipped to be productive. Most people have a favored time of day when they find it easiest to concentrate. I prefer to focus on activities requiring deep work first thing in the morning. I have highly productive colleagues who firmly believe that my early morning schedule is inhumane and instead prefer a Mencken-like evening schedule. Neither approach is right or wrong; they simply reflect personal preferences. You should find the timing that works best for you and then, as much as is possible, arrange your schedule to have the opportunity to pursue deep work accordingly.

Embrace Boredom – Training Yourself to Enter a Distraction-Free State

During the first summer Osamu Shimomura spent working with jellyfish, he tried many different ways to extract the luminescent chemical that was the target of his research. As is common in research, nothing worked. Recollecting that time, he later wrote "I was conceptually exhausted, and could not come up with one further idea". Inspiration came from an unusual source. In Shimomura's words,

> I spent the next several days soul-searching, trying to imagine ... a way
> to extract the luminescent principle. I often meditated on a drifting
> rowboat under the clear summer sky...meditation afloat was safe, but if I
> fell asleep and the boat was carried away by the tidal current, then I had
> to row for a long time to get back to the lab. One afternoon on the boat, a
> thought suddenly struck me--a thought so simple that I should have had
> it much sooner... I immediately went back to the lab.

Shimomura's musings while drifting in his rowboat gave him a hypothesis that rapidly led to a working method that had eluded him in all previous experiments. This method become the core of the years of subsequent work he did on aequorin.

How does Shimomura's nautical tale of inspiration intersect with the concept of deep work? After all, drifting aimlessly in a boat can hardly be described as a "professional activity performed in a distraction-free state". Or can it? A key feature of Shimomura's use of time is that he was, consciously or unconsciously, training himself to be content in a distraction-free state. In other words, he had learned to embrace boredom. We will never know if he would have eventually made the same conceptual breakthrough if he had spent his afternoons updating his Instagram feed (#JellyfishLife #WashingtonSummer #Rowboat), but it seems unlikely.

How well do you cope with boredom? Your gut reaction to the 30-minute experiment I suggested at the beginning of this chapter may be a good indicator of your answer to this question. If you find it difficult to sit quietly for ten minutes or more without feeling compelled to check something on your phone, computer or TV, there is a high likelihood that you will find entering a distraction-free state difficult when it is time to attempt deep work.

I am not arguing that a monastic existence filled with endless hours of boredom is something we should aspire to. The ready availability of books, movies, videos, games and so on can enrich our lives in many ways. If, however, we constantly bombard ourselves with inputs that require minimal attention span and active thinking the habit of craving distraction becomes our brain's default setting. This is the concern voiced by Nicholas Carr we saw at the beginning of the chapter: "I'm not thinking the way I used to think....I get fidgety, lose the thread, begin looking for something else to do".

The suggestion to embrace boredom as a tool to improve deep work comes from Cal Newport, and I heartily agree with it. This strategy reminds us that we not only train ourselves to do deep work when we are actively working but that our leisure and non-work time also has a strong influence on our ability to think clearly. Best of all, adopting this strategy is simplicity itself – just find an enjoyable activity that doesn't present you with an endless array of "shiny objects" and regularly spend time doing it. The aim of doing this is nicely captured in an album title from musician Courtney Barnett, *Sometimes I Sit and Think, and Sometimes I Just Sit.*

Training yourself to detach from the endless series of distractions that surrounds us will, over time, make you more comfortable and equipped for the intense concentration that is required for deep work. It is also likely to improve your general sense of calm and well-being. So instead of viewing boredom as a negative state that must be avoided at all costs, look for ways to embrace boredom as a way to increase your capability to enjoy distraction-free thinking or sitting.

Make Conscious Choices about Technology

There can be no doubt that technology has made doing research easier. When I was doing my PhD in the 1990s, getting technical papers involved walking to the university library, finding journals on the shelves and remembering to take a pocketful of change to use a photocopier. Sending a manuscript to a journal involved carefully printing physical copies of the text and figures to mail to the publisher. This can all be achieved now from my office or back porch with a few keystrokes on my laptop.

Beyond the realm of research, our social connections have been transformed by technology. Professional and social communication by letter or phone call has long been replaced by email, which in turn has been pushed aside by text messaging, Facebook, Twitter, Instagram and Slack. By the time you are reading this, additional platforms that are currently just a glimmer in the eye of a startup founder somewhere will have been added to this list.

Are these technologies good or bad for our capacity to do deep work? To limit the scope of this question even further, let's just think about the example of getting technical papers to read. Online access to technical journals has most certainly been a positive development. Hours that used to be spent fiddling with photocopiers have disappeared, literature searches can be far more efficient and sources can be stored electronically for easy access no matter where you are physically located. But these efficiencies have come at a price. It is easier to download a few more papers to skim than it is to deeply ponder an individual paper, a human tendency encouraged by online journals' colorful graphical abstracts and automated suggestions of related papers. In the old days of walking to the library, the idea of discarding a laboriously photocopied paper without reading it carefully seemed wasteful. The focused searches that online access enables have all but eliminated the serendipity of finding other thought-provoking reading while flipping through a physical journal. From this perspective, online access to technical journals has greatly improved what might be called the shallow work pieces of reading the literature, but it is less obvious whether it has improved the ability of most researchers to complete deep work.

I'm sure you have already realized that the "on the one hand/on the other hand" arguments I just made about reading journal articles can be applied to almost any new technology platform. New technologies typically have real efficiencies and advantages but also have incredible potential to distract us from the sustained concentration needed to do deep work. Let's imagine that many of your friends and family have started using a new platform that we'll call Instabook. The appeal of Instabook seems hard to resist. Many of your friends are already using it. Even your academic department and your favorite technical journals have signed up. When you set up your own account you find that Instabook really is a wonderful way to stay connected to friends and family and to learn about the steady stream of new papers in your technical field. You also quickly find that you can hear from a diverting array of famous and semi-famous commentators. After a few months of regularly using Instabook, you find yourself checking it on your phone while waiting in line at the supermarket and that your morning routine when you come into the office involves spending 20–30 minutes chuckling at the overnight comments on the very active thread associated with your favorite sports team and their rivals. Has Instabook made your life more enjoyable? Quite possibly. Has it improved your ability to do deep work? Probably not.

Your time is a finite resource. Any time you spend scrolling through Instabook or browsing a newspaper online or watching sports on TV is time that is gone forever. None of these diversions are inherently negative; indeed, they bring genuine joy to many people. The quantity of diversions available to us, however, is not finite. The deliberate design of most online platforms, TV stations and video games is to present a never-ending supply of "content". These observations mean that it is vital to make conscious choices about the technology platforms we use if we want to nurture the time and mental focus needed to do deep work.

How should you decide how much to use, or even whether to use, any specific technology platform? The vignette above highlights a typical answer to this question: try out the technology and see what it is like. This approach focuses on the benefits and enjoyment that the new thing can bring you. In Cal Newport's book, he describes a more radical but far more powerful answer to the same question, the 30-day technology fast. This strategy has two parts. First, as the name suggests, stop using a particular technology platform for 30 days. If you can, delete the app from your phone or make some other change that creates a barrier to "forgetting" your resolution and logging in "just for a couple of minutes". Starting a fast on the first day of a month is an easy way to remember when the 30 days is finished, but any day will do if you record it somehow. Second, don't make any grand announcements about your hiatus. Don't change your online status to indicate you are temporarily away and don't announce to other users that you are taking a break. The first few days of this 30-day experience can be difficult if your use of the technology platform you are temporarily foregoing has become a regular habit, but pushing through this initial period is important for the technology fast to be effective.

After 30 days away from Instabook (or whatever technology platform you were evaluating), ponder the following questions. What specific things did you miss out on in the last month because you didn't use the technology? How important are these things to you? Did these things have important long-term implications, or where they short-lived controversies that were replaced by something different a day later? Did other people on the platform miss you (without you telling them you weren't there) or did your absence go unnoticed? What did you do with the time that was freed up by your technology fast? How important are these things to you? Did you notice any change, positive or negative, to how calm you were and your ability to enjoy your life? These questions are intended to help you consider both the positive and negative aspects of the technology platform. If the positives dominate then you have good reasons to go back to Instabook and enjoy using it. You might find, for example, that your family uses Instabook as a primary way to stay in contact and that your ability to receive positive support from them and for you to give support to others is greatly hampered if you avoid Instabook. This would be a wonderful reason to regularly use the technology. Alternatively, you might find that your family and friends mainly use Instabook to share political or other content that does little but raise your blood pressure and promote disagreements. In this case, your 30-day experiment might convince you that finding ways to connect with your family outside of Instabook and skipping all the online vitriol would improve your life.

The aim of a 30-day technology fast is not to train yourself to become a digital ascetic who communicates only via letters written with a feather and ink. Modern technologies can affect our lives in many positive ways, but we need to control them rather than the other way around. A 30-day technology fast is a well-defined strategy to decide whether you should allow a particular technology to take up your precious time and attention. Making these conscious choices about the technologies that you use is closely related to the strategies of establishing schedules for your work and building the capacity to embrace boredom in strengthening your ability to do deep work. Together, these three strategies will give you greater control of how you use your time and help you enjoy the long-term satisfaction of accomplishing deep work.

Chapter Summary

- Deep work is defined by professional activities done in a distraction-free state that push cognitive abilities to their limit. Long-term success in research and similar creative endeavors is strongly connected to the ability to do deep work.

- The ability to do deep work creates a tremendous competitive advantage, particularly in a society where trends of various kinds are diminishing most people's abilities in this area. Deep work is a rare and valuable skill.

- Performing deep work is intensely satisfying. Although deep work often brings long-term external rewards, most people who excel at deep work do so primarily because of the internal satisfaction their work brings.

- Technology and other aspects of life surround us with a near-infinite capacity to be distracted from deep work, so accomplishing deep work requires intention and planning.

- Scheduling the times at which you perform deep work is a powerful strategy for training yourself to set aside the distractions of shallow work and frippery and to overcome the sense that you don't have time for deep work.

- Learning to embrace boredom, not only while working but also at other times, helps nurture the patterns of distraction-free thinking that are a core part of deep work.

- Place a high premium on your time and attention when considering whether to regularly use any particular technology or social media platform. Use a 30-day technology fast to evaluate the advantages and opportunity cost of key platforms you use regularly.

Further Resources for Chapter 2

The structure and ideas of this chapter are based strongly on Cal Newport's excellent book *Deep Work: Rules for Focused Success in a Distracted World* (Grand Central Publishing, 2016). In relation to the choices about technology mentioned at the end of the chapter, it is worth noting that Newport is a successful academic researcher and widely read business author who does not use Twitter, Facebook or any similar social media.

Osamu Shimomura's adventures collecting and processing jellyfish are described in a very entertaining article he wrote 13 years before sharing the Nobel Prize: A Short Story of Aequorin, O. Shimomura, *Biological Bulletin*, 189 (1995) 1.

Georgia Tech's role in early cold fusion experiments, as well as many other scientific misadventures in the nuclear engineering community, is described in James Mahaffey's wonderful book *Atomic*

Adventures: Secret Islands, Forgotten N-Rays, and Isotopic Murder – A Journey Into the Wild World of Nuclear Science (Pegasus Books, 2017). I benefited from and enjoyed conversations with Dr. Mahaffey and with John Toon at Georgia Tech in learning about cold fusion history.

The history of Alfred Gilman and Louis Goodman and their contributions to medical science was drawn from *The Drug Hunters: The Improbable Quest to Discover New Medicines*, Donald R. Kirsch and Ogi Ogas (Arcade, 2016). Goodman's failed experiments with curare were also described in his obituary in *The New York Times* in November 28, 2000.

H. L. Mencken's daily habits, and the habits of many other creative artists and writers, are described in *Daily Rituals: How Artists Work*, Mason Currey (Knopf, 2013).

3

The Seven Habits of Highly Ineffective Researchers

Self-help authors love to make numbered lists. Norman Vincent Peale's *The Power of Positive Thinking* has ten rules for obtaining confidence, three secrets for vigor and five techniques for overcoming defeat. Don Miquel Ruiz titled his book *The Four Agreements*, and Deepak Chopra had a bestseller with *The Seven Spiritual Laws for Success*. Although it is easy to be cynical about these books, one book based around a numbered list has stood the test of time: *The Seven Habits of Highly Effective People* by Stephen R. Covey. Covey's practical insights ("Begin with the end in mind") have genuinely helped many people since his book first appeared in 1989.

In this chapter, we will look at some of the habits that can enhance or inhibit career success in research and related creative pursuits. We will do this by turning Covey's upbeat point of view on its head and considering the habits of highly *ineffective* researchers. The aim of doing this is of course not to encourage you to adopt bad habits. Instead, I hope that you can recognize the tendencies towards these bad habits that all of us have and find ways to avoid them in your own work.

Bad Habit # 1: Believe Your Own Hype

When my children were small, one of them asked at the dinner table what I did at work. I explained that my research group was working on a fuel that could one day replace gasoline as a fuel in cars, since at the time we were studying materials for storing hydrogen in vehicles. My family seemed satisfied with this information and conversation moved on. A few days later, I was asked at dinner "Dad, that fuel you were making, did you get it to work?". My kids had done something that often gets forgotten in discussing research: they checked on a hype-filled description of a research aim to see how it held up under scrutiny.

Hype plays an important role in research. Researchers need to justify and compete for the funds they use to support their work. A proposal that promises to "revolutionize" the tools used for some measurement is more likely to be successful than one that mildly claims that the proposed work will "add incrementally to the understanding of an area in which hundreds of papers have already been written". Universities and research institutes maintain interest from alumni and partners by writing about research "breakthroughs" and "insights".

There is great value in being able to explain your work to a non-technical audience in ways that leave them thinking it sounds interesting and worthwhile instead of being a colossal waste of time and money. But being able to make something sound exciting for someone that can't understand any of the technical details is not the same as actually doing impactful research. Highly ineffective researchers convince themselves that combining a series of topical buzzwords in an abstract really is enough to drive their research forward.

Believing your own hype can be a severe impediment to making real progress in your work. Writing "develop a fuel that can replace gasoline" or "cure cancer" on your daily to-do list doesn't help in the slightest to plan a series of concrete actions you should take. Having ambitious long-range goals is laudable, but only if they are backed up by fine-grained planning. Believing your own hype can also prevent you from being honest about potential shortcomings in your work. It is common for subfields of research to generate a large collection of papers with introductions that confidently claim that improving some physical property would unlock some real-world application. Blithely accepting these claims as true rarely leads to significant research progress. Instead, think through such claims from the point of view of a skeptic who doesn't have a vested interest in the subfield being valuable.

Solving complicated real-world problems is rarely possible by overcoming just one barrier. One helpful antidote to believing scientific hype is to make a list of things that could go wrong on the path from what has actually been done to having the ultimate promise of the hype being realized. This is likely to show that the source you are looking at gives little in terms of the details of the work that is being reported, a situation that is particularly likely if you are looking at a story from a press release or the popular press. Realizing that the details are unclear is a good indication that you should be especially attuned to the possibility of unjustified hype being present.

A useful way to calibrate yourself in terms of your susceptibility to hype is to reflect on who you have in mind as the audience for your work. In other words, who are you aiming to impress with your work? The audience that is most easy to impress are non-technical people who like you personally because of your relationship with them. My mother is a retired English teacher who has always been unfailingly supportive of me. I cannot imagine her being critical of my research, even if one of my group's papers was filled with turgid prose and the occasional grammatical error. I am extremely fortunate to have a loving and supportive family, but it would not be wise for me to base my assessment of the technical quality of my work on my mother's opinions.

A second potential audience you could aim to impress is researchers who work in related but different fields. This is an audience you might encounter while interviewing for a job in a company or university or when writing a grant application. Thinking about the expectations of this audience can be extremely helpful. "Everyone" in your subfield might agree that some particular technical goal is worthwhile, and they would surely agree that the subfield itself is intrinsically important. How would you justify these judgments to other technically accomplished people who have either never heard of your field's goals or are not convinced that the subfield itself is even interesting? Hiring committees and grant reviewers are often react negatively to hype that can't be backed up with genuine depth and understanding.

The third and most important audience you should seek to impress with your work are the people who are fully immersed in the technical details relevant to your work. I once arrived at an international conference, slightly bleary from jet lag. Before I had even found my name badge, a young researcher I had never met started a conversation with me saying something like "In Figure 3 of your paper in JPCC last year, why did you choose framework density as the x axis, not pore diameter?". I doubt that I provided a coherent answer to his question, but his intensity was a good reminder that other researchers are often ready to critique my methods and results. In the hit sitcom "The Big Bang Theory", a key character is Sheldon, a brilliant but socially awkward and obsessive theoretical physicist. If Sheldon sees a problem with someone else's reasoning, he is sure to blurt it out. The show is popular with many scientists because they recognize the behavior of their colleagues in Sheldon's antics. As an antidote to believing your own hype, think about having your field's equivalent of Sheldon sitting in the audience as you give a talk or reviewing your latest paper.

Bad Habit # 2: Don't Learn from the Past

Scientific research is all about moving forward. Doing new things. A review of a manuscript that says "This work reiterates results that have been well known for years" is not going to make a journal editor's heart beat faster. On top of this, the scientific literature moves quickly, with hundreds or thousands of new papers appearing in each subfield each month. To stay at the forefront of this tsunami of information, highly ineffective researchers opt to stay in touch with only the very latest in their field. This choice saves enormous amounts of time by removing any temptation to learn from the past.

The weakness of ignoring the past in your research field is captured in George Santayana's famous aphorism "Those who forget the past are condemned to repeat it". Scientific progress will be slowed down if the hard won insights from a generation of researchers are ignored a few years later. Not being aware of the history of a field puts you at risk of receiving reviews for your journal manuscripts like this one, which I saw as a journal editor: "The topic is very interesting and the paper is clearly written. However, my major concerns are: (1) Novelty of the conclusion is unclear, (2) Data is not rigorously collected, and (3) Analysis and discussion lack depth".

From a purely pragmatic point of view, it is helpful to realize that there are prominent people in your field who have a longer memory of past work than you do. This is often because they were the ones that did the work. It is quite likely that there is at least one researcher in your current department or institute who has been working actively since before you were born. Multiply that individual's experience by many times and you are beginning to grasp the amount of experience that exists in your field. A sure way to make a positive impression on this important group of people is to show genuine interest in their lifetime's worth of work.

I hope I have been able to convince you that ignoring the past in your field is a bad idea. How can you avoid this trap in your own work? A simple approach to this issue is to make sure you include sources in your reading that reach back more than a few years. Many online search methods skew strongly towards the most recent information. This trend is exacerbated by the focus many journals and their readers put on short-term measures like impact factors. As you read the literature in your area, take time to follow chains of references backwards in time. What are the seminal papers in your field from five years ago? Ten years ago? Twenty years or more? If you can't readily answer these questions, ask a colleague or mentor who has been part of the field's history for their opinion. They are likely to be thrilled to have their expertise valued.

The discussion above implicitly centered on learning about a field from the research literature. By their nature, books appear more slowly than research papers and are often intended to give an overview of a field or provide a particular point of view. It is therefore useful to identify the seminal textbooks in your field. It is likely that much of the content in these books is "assumed knowledge" for anyone embedded in the field. Thoughtfully reading these key books is likely to be richly rewarded in terms of the perspective you will gain on your work. (If the idea of concentrating for long enough to read a technical book fills you with dread, it may be useful to go back and re-read Chapter 2 on the topic of deep work.) I was once working on a research project that had stalled for months because no matter what we tried, the computer code we had developed wouldn't work. Fortunately, a world-class expert in the field from a world famous university came to visit my department to give a seminar. He listened patiently while I explained my problem to him, and then calmly said "Well, if you read my book...". Sure enough, the problem that was the root of the months of difficulties in the project was clearly described in his book. I would recommend that you avoid my embarrassing situation by reading the foundational books in your area long before you meet their authors or present your work at a conference they might be attending.

Bad Habit # 3: Work Alone

Major scientific prizes are awarded to individuals or very small groups of research leaders. Jobs and promotions are given to individuals, not to teams. One of the greatest indignities in research is to be "scooped" by a competing researcher who reports a breakthrough discovery that is almost identical to work you have been pursuing for months or years. A highly ineffective researcher surveys these facts and comes to an inescapable conclusion: it is better to work alone, not sharing ideas with others and ensuring that when credit for work is given, there is only one person to whom it can be assigned.

The mathematician Andrew Wiles may seem to be a living advertisement for the power of working alone. Wiles spent six years working by himself at home on proving Fermat's Last Theorem, one of the most famous results in mathematics. He was so secretive about this endeavor that only his wife knew what Wiles was working on. Before announcing his complex proof to the world, Wiles asked his Princeton University colleague Nick Katz to help him verify the logic of his work. Knowing that this process would take many hours of working together, Wiles and Katz came up with an ingenious plan to avoid tipping off their colleagues that something unusual was going on. Wiles announced he was teaching a new graduate-level course with a nondescript title. On the first day of class, Katz sat in the classroom and Wiles simply started working through his proof without any explanation of where it was leading. After a few class sessions, all of the befuddled students who had signed up for the class had dropped out, leaving only Katz sitting in the classroom working with Wiles. After completing this process, Wiles announced his proof of Fermat's Last Theorem at a conference to worldwide acclaim.

The proof of Fermat's Last Theorem has an amazing postscript that also involves Nick Katz. Wiles submitted his work to a top journal, which sent it to several referees, including Nick Katz. Katz later said he did nothing for several months but work on checking the details of the manuscript. After much work and discussion with Wiles, Katz concluded that the proof was incomplete. Wiles had not proved the theorem! This devastating realization sent Wiles back to work again, and for a year, he and a former student who had worked with him, Richard Taylor, tried to plug the gap in his original proof. Finally, Wiles realized that a technique he had tried intensively three years earlier but then abandoned was just what he needed to solve his problem. Fermat's Last Theorem had indeed been proven. Although Andrew Wiles name will forever be associated with this proof, Nick Katz and Richard Taylor played key supporting roles.

A prime reason to not work entirely alone is that no matter how smart you are, it is incredibly unlikely that you have a monopoly on good ideas. Talking about your data, your hypotheses or the barriers that are stopping you from moving forward with others can have a range of positive effects. Most obviously, the person you are talking to might have a good idea that helps you. I have sat in many thesis committee meetings and research seminars where a question from someone peripheral to the work being discussed brought up a new creative idea. Less obviously, telling someone else about your work means that you have to gather your thoughts and tighten your logic. It is common for university faculty to say they didn't really understand some particular subject that is foundational to their field until they had to teach it. Teaching something to a room full of inquisitive students is a powerful way to detect assumptions the instructor had previously taken for granted. The act of putting your thoughts into words, either in person or in writing, forces you to decide what is important and what is not. This is exactly what Andrew Wiles did in sharing his work with Nick Katz, first in the form of a presentation in a classroom and then in the more formal structure of a written manuscript.

A final and probably more powerful argument for not working alone is that sharing the frustrations and small victories of your research with colleagues or collaborators will make your work more satisfying. This observation is closely related to our discussion of deep work in Chapter 2. Finding ways to exchange ideas with others is, like accomplishing deep work, something that will make you feel good. My choice of "exchange" here is deliberate. If you have an acquaintance who talks incessantly about themselves but never has any interest in your life, it is unlikely you will want to develop a lasting friendship. To avoid acting like this tiresome acquaintance yourself as you talk with your colleagues, find ways to actively learn about the work they are doing as well as talking about your own work.

Bad Habit # 4: Confuse Activity with Productivity

Doing creative work takes time, and lots of it. If 40 hours of work a week in the lab can ensure progress, then 60 hours will mean 50% more progress. Taking a day off on the weekend or, even worse, an extended holiday away from work can only subtract from progress. It is also important that the people around you, particularly your boss, know when you are working hard. What is the point being in the office late if the people who went home "early" don't know about it?

The thinking of a highly ineffective researcher I have just outlined equates time at work with research progress. This kind of thinking implies that the world can be neatly divided into work and non-work and that less of the latter means that more work gets done. If you worked at an assembly line making car parts, this kind of view would be justified. But as we have already discussed in Chapter 2, in the realm of creative work not all tasks have equal long-term value. Highly ineffective researchers ignore this nuance and confuse activity with productivity. An ineffective researcher can be at work a lot but achieve only a little.

Covey's book *The Seven Habits of Highly Effective People* describes two extremely helpful ideas for time management that can help prevent confusing activity and productivity. First, he suggests categorizing demands on your time as urgent or not urgent and, separately, as important or not important. Of these two categories, urgency is the easiest to understand: anything that demands your attention right now is urgent. Finding out why the fire alarm in your building is ringing is urgent, but posting online photos of your pet is not, no matter how cute they are. The distinction between important and not important can be fuzzier, but most examples are easy to categorize. Thinking deeply about your data is important. Putting the collection of books in your lab into alphabetical order is not important.

The categories above point to two groups of activities that require special attention. If something is urgent and important, then you have little choice but to prioritize it. A reminder from a program manager at your funding agency that they need to receive a report on your grant by Friday to approve funding for the next year's work should prompt immediate action.[1] The most significant group of activities, however, is typically those that are important but not urgent. These are the long-term deep work activities that are at the core of making real progress in creative research. Many of the items that fall into the urgent and important category are shallow work, using the terminology from Chapter 2. Think about the famous researchers we have encountered so far, for example, Andrew Wiles, Osamu Shimomura and Marie Curie. None of these individuals are remembered for work they did that was urgent. Instead, they devoted large blocks of time to work that they considered important, consistently prioritizing this above items they decided were not important or merely urgent.

The urgent/not urgent and important/not important categories are a great start in knowing what to work on, but they don't give much guidance about how to actually organize your time. To tackle this issue, Covey gave a striking illustration. Imagine that your time in the upcoming week is represented by the space in a large jar. No matter what you try there is no way

---

[1] This real example from my email inbox also illustrates why many "urgent" items only attain this status because of personal choices about time management. The need to write the report mentioned in this example had been known to me for months. By procrastinating in writing the report, it moved from something that was not urgent to something that was truly urgent.

to find more hours in a week, and the same is true for the volume of the jar, which is fixed. The various ways you could use your time are represented by rocks of different sizes: large rocks for the tasks you classified as important, small pebbles for urgent but not important items, and fine-grained sand for items that are neither urgent nor important. Having assembled your rocks and sand, you are ready to plan your week by filling the jar.

One approach to filling the jar is first to pour in all the sand and then add a layer of pebbles. In time planning terms, this fills in a significant fraction of your time (the jar) and removes many individual items from your to-do list (the piles of rocks and sand). Unfortunately, when you try to add a few larger rocks, there is little room left. This scheduling approach strongly limits your ability to work on those vital important but not urgent tasks. An alternative strategy is to put the larger rocks in the jar first; that is, set aside time for the important but non-urgent tasks before allocating time to anything else. Once those rocks are in the jar, pebbles can be added and, with a little gentle shaking, they find ample spaces between the large rocks. Finally, sand can be poured into the jar, and it too will find spaces between the rocks and pebbles.

I hope the message from Covey's illustration is clear. Planning your time around activities with long-term value takes discipline, but hoping that time for these tasks will emerge spontaneously among the many demands on your time is foolhardy. A highly ineffective researcher measures their progress by the total number of hours they have spent in the office or lab. A highly effective researcher pays closer attention to how they are using their time, knowing that even small increases in the periods they spend on focused deep work have greater long-term impact than long hours of poorly focused busy work.

Bad Habit # 5: Exclude All Inputs That Aren't Immediately Relevant

Reading the research literature can be a never-ending pursuit. At many large research institutions, there are seminars every day of the week that can provide coffee, snacks and distraction to young researchers. Well aware that these kinds of activities loosely connected to research can expand to fill all their available time; a highly ineffective researcher places strict limits on the topics to which they will devote mental energy. If a source of information isn't immediately relevant to their current research project, the ineffective researcher reasons, then learning about it is detracting from getting that project finished.

There is a grain of truth in the thinking I just outlined; if every available hour of your time is filled without making tangible contributions to your specific research project, then your productivity will stall. The essence of progress in research, however, often requires creative insights. Where can these ideas come from? Chapter 4 is devoted to systematic ways to generate creative ideas, but until we get to that discussion, it is enough to point out that many creative breakthroughs rely on combinations of previously known concepts. As W. I. B. Beveridge wrote in *The Art of Scientific Investigation*, "Originality consists of linking up ideas whose connection was not previously suspected". The same sentiment was put into a pithier form by Steve Jobs, who said "Creativity is just connecting things".

If Beveridge and Jobs are even mostly correct, and I think they are, then where can you get the ideas and things to connect to make creative advances? The ineffective researcher who rigorously excludes all inputs without obvious and immediate relevance may boost their short-term productivity but at great cost to their long-term creativity. I suggest that you survey the sources that can give you research-related ideas over an extended period, for example, the next year. Most technical fields have one or more magazines written for a broad technical audience that follow trends and advances. Reading these sources regularly is one useful source of technical information. Attending a regular seminar series, if your institution has one, or finding an online substitute is a useful way to be exposed to a continuing string of new ideas. When I look at my own weekly schedule, the temptation to skip seminars that have topics well outside my specific research interests can be strong, but it is likely that these are the talks from which I can learn the most. Reading and going to seminars in this way can both be accomplished in a couple of hours each week. It is also well worth finding a conference or workshop to attend with the specific goal of expanding your knowledge once a year or so. If you go to a mega-conference with hundreds of parallel sessions, devote half a day or a day to sitting in sessions in a field outside of your own. Alternatively, go to a smaller workshop in a field intellectually adjacent to your own. For seasoned researchers, a useful metric for these events is how many of the speakers and attendees you already know: if you already know most of the people around you, then the scope for learning genuinely new things may be limited.

Putting yourself in environments like seminars or conferences could, I suppose, lead you to new ideas by some sort of creative osmosis. In my experience, however, a more active approach is needed. After listening to a talk or reading an article, it is helpful to try and answer some or all of the following questions. What was the key technical advance? Why does it matter? Does it actually matter? What was the key technique the work used? Could I explain how that technique works, even in hand waving terms?

The questions I just listed can help synthesize the ideas you have heard or read into a form you might remember. There is one more question to ponder that directly addresses the concept of creativity as "just connecting things": is there an analogy between this work and my own research? At first, the idea of finding these analogies can feel strained and perhaps even impossible. With repeated practice, however, the simple question can become a catalyst for new ideas. I once spoke on this topic to a group of engineering PhD students and decided to spice up my talk with a competition. I introduced the group to the musical concept of a hemiola, a rhythm in which a group of six notes can be heard either as two groups of three or as three groups of two,[2] and offered a modest cash prize for the best analogy that could be drawn between this idea and an engineering concept. (Can you devise a solution to this challenge?) The blank looks on the audiences' faces made me worry that my example would fall flat, especially since I couldn't think of any analogies that fit the bill myself. To my delight, the students came up with multiple suggestions, including the six-stroke engine, a design that extends the common four-stroke internal combustion engine cycle with pistons that take three strokes in place of two. I am not claiming that the six-stroke cycle was invented on the basis of inspiration from classical music theory. Nevertheless, it is interesting to realize that once the hemiola/combustion engine analogy has been conceived, it seems in retrospect almost natural to explore extensions of the four-cycle engine to other "rhythms". We will come back to the challenge of using ideas you see elsewhere to generate new approaches in your own work in Chapter 4.

Bad Habit # 6: Criticize Others but Never Yourself

The ability to make judgments about the work of others is a vital skill in any area of research. Peer reviewers for scientific journals are expected to offer an opinion on whether the work in a paper they are reviewing is correct and, if it is, whether it is important. Reviewers of grant proposals have to answer the same question but in a context where only a small fraction of proposals can be rewarded with funding. Not everything that gets published in scientific journals is completely correct, so researchers are trained to maintain a healthy skepticism as they read the literature. Highly ineffective researchers are experts at criticizing the work they read and hear about, but they also understand that applying the same demanding standards to their own work is tiresome and unfair. After all, how could all those control experiments be completely documented when they were assigned to an undergraduate researcher who was also searching for a job and about to graduate? And putting a balanced list of literature references in the introduction to that manuscript would have taken so much longer than the time that was available.

The tendency to have different standards for other people and for ourselves goes back far in history. In the biblical book of Matthew, Jesus rhetorically

---

[2] The opening theme in the first movement of Joaquin Rodrigo's beautiful *Concerto de Aranjuez* for guitar and orchestra is one famous example of a hemiola.

asks a crowd "Why do you look at the speck of sawdust in your brother's eye and pay no attention to the plank in your own eye?". Jesus wasn't talking about doing high-quality research, but his description of a habit of highly ineffective researchers is apt. The examples listed above illustrate part of the issue; when we think about something we did ourselves, we know the context in which the work was done and can easily construct a narrative that justifies our approach. In contrast, when we see work by others, the "excuses" that might become part of a justifying narrative aren't available.

The remedy to this issue is to become self-critical about your own work. When is a good time to employ self-criticism? When the results of your work look bad, to start with. If your experiments won't work or your writing seems muddled, deliberately taking on the role of a critical reviewer and listing potential problems can be a useful way to move towards solutions. Another time to exercise self-criticism is on the days when your results look especially good. Any time your results turn out just the way you hoped is a good time to check your assumptions. If you ever find yourself thinking that some results are "better than we could have hoped for" or "too good to be true", take a deep breath and think about what a highly combative reviewer would say about them. (The Cold Fusion debacle detailed in Chapter 2 is an extreme example of this advice being ignored.) Finally, any time you aren't getting any results at all is also a good time for some self-criticism; if independent reviewers were given the job of assessing your current approach, what would they find? In short, self-criticism is appropriate no matter how your work is going.

It is important for me to be clear about what I mean by self-criticism. Self-criticism does *not* mean denigrating your own abilities and telling yourself "I am a useless worm and nothing I touch will ever turn out right". If your internal dialogue ever starts to sound remotely like this, please find someone to talk to about your mental health! In a research context, healthy self-criticism looks at your work in as close to an unbiased way as possible, anticipating problems that might occur later to others. Healthy self-criticism acknowledges that doing creative work is hard, and things rarely turn out exactly right the first time around. This kind of attitude agrees that taking a step backwards might be required before forward progress can be made and that getting things right is more important than moving faster.

How can you encourage a positive culture in which self-criticism (and criticism by others) leads to better creative research? First, and most crucially, always critique the work, never the person. The tone and wording used to deliver criticism, even to yourself, matter a great deal. Second, when looking over your own work or presenting it to others, have an attitude that genuinely welcomes questions. Treat each question as a chance to potentially learn something new instead of being defensive. Third, don't be satisfied with superficial answers, especially your own. Most researchers have had an experience where they give a talk and time constraints mean that they must answer a question in just a few seconds before the next speaker needs to start. Instead of answering every question like you were in this situation, imagine instead that your questioner would listen patiently for as long as it took to give a nuanced, thoughtful answer.

Highly inefficient researchers readily see the metaphorical sawdust in the eye of everyone around them but makes themselves comfortable by ignoring the plank in their own eye. Instead of falling into this all too human tendency, develop the ability to regularly apply constructive self-criticism to your work as a tool to sharpen your thinking and unlock new ideas.

Bad Habit # 7: Let Your Personal Background Hold You Back

Each of us is shaped by our personal history and the family in which we grew up. Even the most loving families can be complicated, so almost everyone can think of some ways in which other people they know seem to have an easier life. In Habit # 6, we looked at how a self-narrative can often be invented to justify potential shortcomings in creative work. Highly ineffective researchers take this skill for self-narrative to greater heights by applying it to their personal background and finding reasons to limit their current state. After all, the field is filled with people who have better preparation and education, not to mention better equipment, more funding, more creative colleagues and probably less complex personal lives. How can an ineffective researcher be expected to succeed in comparison?

To illustrate the choices that exist in self-narrative, we will look a little at my background. I grew up in a small country town in Australia called Armidale, with a population of about 20,000. In many ways, it was a fantastic place to grow up, partly because it was disconnected from the problems of the outside world. Armidale was, and still is, surrounded by bushland in every direction. The nearest town of a similar size is over 100 km away and that town's main claim to fame is that it is the "country music capital of Australia". When I was a kid, there were two TV channels – if you didn't like what was on one you could always try the other. Armidale is connected to Sydney, Australia's largest city, by train. Around the time I was born, the state railway decided it was time to move into the modern era by replacing the steam train they used with a diesel one. If I wanted to convince myself that my background could hold me back, this description of my home town would be a good starting point.

Around the time I was finishing high school in Armidale, a young man and his mother moved to town from the Northern Territory, which is considered remote even by Australian standards. When he was seven, the boy had been kicked in the head by a horse and was put in a coma for a week before recovering. Despite this early misadventure, he pursued typical teen-age interests such as riding skateboards and bicycles. It turned out that he was really good at cycling and he went on to become a professional cyclist. In 2011, Cadel Evans won the Tour de France at the age of 34 after coming second in the same race in 2007 and 2008. Not bad for a boy from a small town in the Australian bush.

Many years earlier, another young boy attended primary school in Armidale. John Cornforth was then sent to a high school in Sydney before attending Sydney University, where he studied chemistry. In 1939, just as World War II began, Cornforth moved to Oxford University to continue his studies, in part because at that time no university in Australia offered a PhD in chemistry. After receiving his degree, he remained in the UK, doing research on the synthesis of biologically relevant molecules like cholesterol. This work led to Cornforth sharing the Nobel Prize in Chemistry in 1975. There is one more aspect of John Cornforth's life that make his accomplishments even more remarkable. When he was a teenager, Cornforth was diagnosed with an ailment that progressively damaged his hearing. By the time he was an undergraduate, he was completely deaf.

The amazing successes of Cadel Evans and John Cornforth support the narrative that people from my home town can reach the top of their chosen fields, even if it means overcoming some significant hurdles along the way. Does this mean that my background will inevitably lead to the same level of success? Of course not. The achievements of Evans and Cornforth are so notable because they are unusual, not because everyone from the town was similarly talented. If we reject this success-inducing background description, however, then we must also throw away the idea that my personal trajectory was automatically limited by the remote nature of my home town. We are each shaped by our personal history, but that history does not make our future inevitable.

I hope that many readers will be able to look back on their lives and see many positive aspects of their life history. I am hugely fortunate to have grown up in a loving home with parents who were highly educated; my father was a physicist at the local university, and my mother was a school teacher and administrator. I have even had the good fortune to write a technical paper with my Dad, something that not too many families can claim. Some readers, however, may have genuinely horrible experiences in their past, such as overt discrimination, dire financial hardship or instances of physical or mental illness. Significant challenges like these have long-lasting implications, and I am not suggesting that you just ignore them and "think happy thoughts". Indeed, these kind of experiences may be a good reason to find professional help from someone who can help you process them. For most people, though, life contains a mixture of wonderful and not so wonderful events, leaving a choice about how to think about the connection between the past and future. Effective researchers are able to view their background in a balanced way, realizing that the people around them have also benefitted from privileges and had to deal with adversity.

Chapter Summary

We have explored seven bad habits that may seem natural or even positive in the short term that ultimately limit a researcher's ability to work effectively. The summary below restates these bad habits in terms of positive habits you can establish to increase the effectiveness of your creative work.

- Maintain a healthy skepticism about glib explanations of the significance of your work (or the work of others). Don't fall into the trap of believing your own hype.

- You can benefit greatly from the history of your field, which extends many years into the past beyond when you began your studies. Take time to identify and read seminal historical papers and books in your field.

- Seeking input from others is a powerful way to clarify your own thinking and to get new perspective on your work. Don't insulate yourself from others by assuming that you need to work alone.

- When you assess the progress you have made, focus on productivity rather than simply on activity. There is no doubt that progress in creative research takes time, but learning to distinguish between deep and shallow aspects of time spent working will boost your productivity.

- Creativity in research frequently relies on combining disparate ideas in new ways. Be deliberate about regularly placing yourself in environments that can expose you to ideas outside the short-term scope of your work.

- Detecting problems in your work through self-criticism is far less painful than having problems identified by reviewers of your papers or grant proposals. Healthy self-criticism is a mindset that applies the same high standards to your own work that you apply to the work of others.

- Find aspects of your personal background that you can be proud of and thankful for, and choose to focus on them whenever you are tempted to compare yourself with others.

Additional Resources for Chapter 3

It is easy to poke fun at self-help books, which are replete with subtitles like "Be You, Only Better" and "Master Your Mind and Defy the Odds", but Stephen R. Covey's *The Seven Habits of Highly Effective People* (published as a 25th anniversary edition by Simon and Schuster in 2013) is genuinely terrific.

The research project I was involved in that was ultimately successful because of reading the right textbook is described in Brownian Dynamics Simulation of the Motion of a Rigid Sphere in a Viscous Fluid Very Near a Wall, D. S. Sholl, M. K. Fenwick, E. Atman and D. C. Prieve, *Journal of Chemical Physics*, 113 (2000) 9268–9278.

The story of Andrew Wiles' proof of Fermat's Last Theorem is described in a documentary for the BBC TV show Horizons that is available online from a variety of sources and also in the book *Fermat's Last Theorem*, Simon Singh (Harper Collins, 2002).

The paper I wrote in collaboration with my father is Diffusion of Hydrogen in Cubic Laves Phase $HfTi_2M_x$, B. Bhatia, X. J. Luo, C. A. Sholl and D. S. Sholl, *Journal of Physics Condensed Matter*, 16 (2004) 8891–8903.

4

How to Have Outrageously Good Ideas

Imagine two of your colleagues are talking about your accomplishments, perhaps in the process of nominating you for an award. "She works so hard", one says, "she just won't give up". "Sure, but the thing I really admire is how creative she is – she just seems to repeatedly come up with outrageously good ideas" replies the other. Which of these two compliments would you find more satisfying? Receiving any compliment feels good of course, but being viewed as creative has special cachet. After all, anyone can work hard, but being creative feels like a rare and valuable gift. In careers focused on research, creativity separates superstars from the rest of their field.

Outrageously good ideas are usually easy to identify in retrospect. Some retain some mystery even after careful examination. Beethoven's symphonies, Einstein's theory of relativity and Pink Floyd's 1973 album *The Dark Side of the Moon* each dramatically changed the history of their field and dramatically diverged from earlier work. Others are so simple that once you hear them, it is hard to believe you hadn't thought of them before. Consider, for example, the humble banana. Children around the world learn to peel bananas by twisting the stem to break the skin and then pulling the skin off the fruit. It might surprise you to know that this isn't how monkeys open a banana. Many monkeys will hold a banana's stem in one hand and gently pinch the skin at the banana's other end. This opens the banana's skin very easily and at the same time provides a natural "handle" to hold while enjoying the fruit. Try it for yourself – it is quite likely that you will conclude you have been eating bananas the wrong way for your entire life! Once you have been introduced to this alternative for peeling a banana, it seems so obvious that you will wonder why every child (and adult) in the world doesn't already know about it. If ideas like this are so obvious, at least in retrospect, why does finding them seem so unusual?

The creative flashes of insight associated with new ideas seem mysterious, not something that can be organized and scheduled. A famous Sidney Harris cartoon shows two rumpled scientists staring at a blackboard where two blocks of mathematical equations are separated by the words "Then a miracle occurs…". The sudden appearance of a new idea can feel almost miraculous; one moment it isn't there and the next moment it is. As a result, it may seem ludicrous to claim that you can learn to systematically have good ideas. My immodest claim, however, is that this chapter will achieve exactly that – it will give you systematic methods that will boost your creativity and help you routinely generate outrageously good ideas.

The main ideas explored in the rest of the chapter are adapted from the book *Why Not? How to Use Everyday Ingenuity to Solve Problems Big and Small* by Barry Nalebuff and Ian Ayres. Nalebuff and Ayres, who were professors of economics and law at Yale University, wrote about generating ideas for better businesses and government policies, but their strategies also apply amazingly well to research and other creative pursuits. A key message from *Why Not?* is to use structured approaches to asking questions about problems as a springboard for generating new ideas. As we will see below, these are approaches that can be internalized and improved over time. Having great ideas isn't a mystical talent; it is something that can be learned.

Nalebuff and Ayres aren't just ivory tower professors who write about the abstract theory of creativity. They have both put their creativity to work in the marketplace. Ayres, as well as being a widely cited legal scholar, cofounded a tech company called StickK where users sign a binding contract to send money to someone, often a charity or cause they vehemently oppose, if they don't follow through on a personal commitment like quitting smoking or exercising more frequently. Nalebuff, working with a former student, founded Honest Tea, a company focused on making slightly sweetened organic iced teas.[1] The idea of an iced tea that isn't completely filled with sugar may seem like peeling a banana from the other end – completely obvious in retrospect. When Honest Tea began in 1998, however, there was nothing like their product available. Honest Tea was bought by Coca Cola for about $110 million in 2011. If the tools Nalebuff used help him create a $110 million business for something as mundane as iced tea, perhaps they can also help you with your research problems.

The four strategies from Nalebuff and Ayres we will explore for generating ideas have names that are designed to help you remember them: What would Croesus do? Can you tell me where it hurts? Where else would that work? and Would flipping it help? Let's jump into the first one and learn how to generate outrageously good ideas.

What Would Croesus Do?

Croesus was a king in ancient Greece who lived about 2600 years ago. Today, he is primarily remembered for being incredibly rich. He is thought to be the first person to introduce solid gold coins, something he wouldn't have done

---

[1] For the benefit of non-American readers, iced tea is, just as it sounds, cold tea. Although readers from the United Kingdom may think this sounds repugnant, it is extremely common in the USA. Traditionally, iced tea is served either "sweet", meaning with an enormous amount of sugar, or "unsweet", meaning with no sugar. The idea for Nalebuff's company came from the simple observation that it is quite common for people to get unsweet tea and then add a small amount of sweetener.

unless he had access to a lot of gold. His name is now used as a synonym for wealth, so an ultra-rich person is sometimes said to be "as rich as Croesus". Nalebuff and Ayres' first strategy for generating new ideas is to ask *What would Croesus do?*[2] By this, they mean how would we solve our problem if we had unlimited resources? No one actually has unlimited resources, but asking this question is a useful way to better articulate what "the problem" is that you are trying to solve and to come up with potential solutions that can be adapted to a resource-constrained reality.

As a 20th-century example of this approach, Nalebuff and Ayres point to Howard Hughes, a fabulously wealthy industrialist who died in 1976. In the 1920s and 1930s, Hughes faced a problem that still seems familiar today: how could he watch movies he wanted to watch when he wanted to watch them? Hughes first tackled the issue of making sure there were movies that he wanted to watch – by buying a major movie studio and making them himself! To give just one example, Hughes spent millions of dollars, a huge amount of money at the time, directing and producing the 1930 film *Hell's Angels*, an "air spectacle" that leaned heavily on another of Hughes' passions, aviation.

But how could Hughes easily watch movies when he wanted to? Decades before Netflix or even the videotape, there was no simple solution at hand. The problem was exacerbated by Hughes' reclusive and obsessive habits, which meant that there was no chance he could do something like go out to a cinema. While living in Las Vegas in the late 1960s in a hotel, he had purchased because he wouldn't leave, Hughes bought a local TV station.[3] This solved the challenge of watching movies because Hughes could call the director of the TV station and simply tell him to put a movie on! Other more normal residents of Las Vegas probably found the broadcast schedule of the station perplexing, but that wasn't Howard Hughes' problem and he was able to watch a showing of *Hell's Angels* whenever he wanted to.

Howard Hughes approach to watching movies is a great example of a "Croesus solution"; it was wildly expensive and impractical to implement at any meaningful scale. At the same time, it shows that the underlying problem is not impossible. It is hard to be motivated to solve a problem if you are convinced that the problem cannot be solved, so knowing that there is a solution, even if it is impractical, can be a great start.

Let's apply Croesus-like thinking to a different problem. Throughout recorded history, people have enjoyed eating honey, but obtaining honey from a nest filled with thousands of potentially angry bees can be unpleasant. Imagine that you are wealthy food-loving individual and that you want to serve incredibly fresh honey with your meals. Of course, you could delegate

---

[2] This question is a deliberate pun on the motto "What would Jesus do?" that achieved great popularity among young Christians in the 1990s and was originally coined in a book by Charles Sheldon in the late 1800s.

[3] He also bought a nearby casino, so he could move its prominent neon sign, which was apparently visible from his room and kept him awake at night.

the task of honey extraction to other people, and this is essentially what we all do when we buy honey at the supermarket. But you don't want honey that has been bottled at some far away location; you want honey that has come fresh from the hive just moments before you eat it. What could you do?

Since resources are no barrier to solving your honey problem, you might begin by buying some beehives and watching as your personal beekeepers extract honey from them. For the beekeepers, this involves putting on a protective suit, opening a hive and pulling out a comb from which the honey is drained. Not surprisingly, the bees don't find this process pleasing, so the beekeeper is surrounded by a swarm of unhappy bees. Maybe by spending enough money, you could design some kind of spacesuit or airlock so the beekeepers could reach inside the hive and get the honey without irritating the bees?

The beekeeper-as-astronaut approach doesn't sound simple, and it certainly wouldn't be cheap, but it helps refine what the true problem is: opening the beehive to get the honey out irritates the bees. By looking at the problem this way, an Australian father and son team developed a wonderful solution called a flow hive. The interior of a flow hive remains sealed except for the opening through which the bees leave and enter. When it is time to remove honey, the owner simply turns a small tap on the outside, separating some mechanical components inside the hive and allowing honey to flow out through the tap. Promotional videos for the flow hive show smiling adults and children happily collecting honey from the tap without wearing any kind of protective gear. Since first being introduced in 2015, tens of thousands of flow hives have been sold, allowing people around the world to enjoy fresh honey without the accompanying clouds of angry bees. The flow hive is one of those "other end of the banana" ideas; once you see it in action, you have to wonder why it wasn't developed sooner.

The essence of asking what Croesus would do is to come up with a solution (or solutions) to a problem without being encumbered by worries about cost or practicality. As the example with Howard Hughes shows, it can also remove any constraints that come from thinking about how other people might react to the solution. The ideas that you generate from this approach may not immediately solve your problem in the budget- and time-restricted real world, but these ideas can be a remarkable catalyst to viewing your problem from a fresh point of view.

Can You Tell Me Where It Hurts?

Since 2002, Dr. Lisa Sanders from the Yale School of Medicine has been writing a column for the *New York Times* called *Diagnosis* with titles like "Why did this man lose his memory?" and "Why couldn't this man stop hiccupping?".

Each column describes an individual with an unusual or alarming set of medical symptoms and the challenges they went through to have their condition diagnosed correctly. In 2019, Netflix released a series also called *Diagnosis* in which Dr. Sanders uses crowd-sourced ideas to diagnose a variety of rare medical afflictions.[4]

Dr. Sanders' columns again and again give examples where superficial diagnoses were wrong. Almost always, medical relief is only possible after a physician has a flash of insight into the hidden origin of the patient's symptoms, typically by combining information from multiple medical tests and their previous experience. This situation has a lot in common with the broader challenge of generating good ideas. If some obvious solution could be applied to the problem you are working on then it would cease to be a problem. The essence of a good idea is often that others who have considered the same problem haven't been able to create a solution. This analogy between medical diagnoses and problem-solving motivates Nalebuff and Ayres' second strategy for generating great ideas: asking "Can you tell me where it hurts?".

Asking where it "hurts" is shorthand for using probing questions to frame a problem more precisely. In Nalebuff and Ayres' book, they suggest using this approach to generate business ideas, since finding a market need (something that "hurts") and then a product to fill that need is vital to any new business. In the Honest Tea example mentioned above, Nalebuff and his business partner realized that lots of people liked drinking slightly sweetened iced tea but felt "pain" because they couldn't buy these beverages at their local supermarket. The idea of understanding the "pain points" for potential customers has become a key part of the widely acclaimed lean startup model of business development.

The value of refining the real source of a problem goes well beyond the realm of business entrepreneurship. To illustrate this idea, let's consider a situation that is common in many technical settings, namely, that a lot of research seminars and conference talks are boring. Since giving these kinds of talks is a key element of building a successful research career, you presumably have a vested interest in not boring your audience when giving presentations of your own. Let's apply the "tell me where it hurts" strategy to the problem of giving non-boring talks. Just telling yourself not to be boring is extremely unlikely to make your talks better. Instead, it is helpful to probe why some, or perhaps even most, technical talks are boring.

Maybe the whole format of a scientific talk, with the audience sitting in rows watching the talk unfold, is set up to predispose the audience to boredom. If this were true, we could explore ways to change the configuration of the room or the format of the talk. Thousands of people, however, regularly pay money to sit and watch tennis, football and other sports events for long

---

[4] Watching this series is an excellent way to gain a greater appreciation for your own relatively good health and other advantages you have in life.

periods of time. This means that the source of the problem can't be just asking people to be part of an audience. I'm sure you have already seen a problem with my sports-watching analogy. At a sports event, audience members are free to leap out of their seats and yell in celebration or outrage. I haven't seen many scientific seminars like that. Nevertheless, the main source of "pain" can't be that we ask audiences in a talk to be spectators.

Perhaps the source of "pain" is asking the audience to sit silently in a room with dimmed lights for the duration of a talk. After all, the situation I just described sounds like perfect conditions for taking a nap. But you have probably already realized that there are places where people happily pay money to sit in just this kind of environment without falling asleep – as audiences watching a movie in a cinema. Now we are making some progress. We have identified situations where people sit quietly in a dim room and are often bored (scientific talks) and where people voluntarily sit quietly in a dim room and are happy (the cinema).

Our analysis points to a tentative strategy for making your next talk less boring: make it more like a movie. Some aspects of movies are clearly impractical for giving a technical talk. I can't see a way to work a car chase or elaborate special effects into my next seminar talk, for example. A good movie, however, has a compelling plot with a beginning, a middle and an end plus perhaps a few surprise plot twists along the way. Would you describe the latest technical talk you saw in these terms? A talk that is deliberately constructed to tell a story is far more likely to keep the audience interested than a dry recitation of facts and figures.

A reasonable objection to the comparison of a movie and a talk is that movies are edited together from many "takes" that give actors and directors multiple opportunities to perfect their performance and presentation. When you are giving a talk in front of an audience, you only get one take. This observation brings to mind another venue where paying audiences are happy to sit quietly in the dark, namely, live theater. What can you learn from live theater that could make your next talk less boring? In a theater production, the cast members have spent many hours rehearsing each scene. Crucially, the judgment of what works best in a scene isn't left to the actors on stage; instead, it is made by the director who sits in the seats of the future audience. This immediately suggests that if you are preparing for an important talk, you should find a way to listen to the talk from the point of view of your audience. An especially useful way to do this is to make a video of yourself giving the entire talk and then sit still and watch the entire video. Doing this is not fun – you will likely cringe at the way your voice sounds, pick up on verbal tics and space fillers that you didn't know you had and so on. But if you aren't willing to sit through the presentation, can you really expect your audience to be excited about it?

There are many skills involved in giving compelling talks, a topic we will cover in more depth in the Chapter 6. Our aim here has just been to see how addressing the question "Can you tell me where it hurts?" to the problem of boring talks can rapidly lead to some actionable ideas. Give this strategy a try for a technical (or non-technical) problem you are facing, and don't be satisfied with the simplest superficial answers. Just like the success stories described in Lisa Sanders' newspaper column, by continuing to refine your diagnosis of where "pain" is coming from you will refine your understanding of what the true problem is and greatly increase your chances of finding an innovative solution.

Where Else Would That Work?

Pablo Picasso, a giant of 20th-century art, said "Good artists copy, great artists steal". This refreshingly honest appraisal of artistic creativity leads us to our third strategy for generating good ideas, asking *Where Else Would That Work*? The aim of this approach is to find ideas that have been effective elsewhere and adapt those ideas to our own problems.

The core of academic mathematics is proving theorems (*If P, then Q*). Mathematicians learn a standard repertoire of tools for proving theorems, including proof by contradiction (*Assume P and not Q, then show a contradiction follows*), proof by induction, proof by contrapositive and the dauntingly named proof by exhaustion. Having this list of methods is helpful to a mathematician because it offers multiple independent ways to attack a recalcitrant problem. This approach of systematically cataloging potential solutions to problems is also valuable in other creative endeavors. Once you develop the habit of asking *Where Else Would That Work*? when you see a clever idea, you may find that collecting a diverse set of problem-solving techniques is almost irresistible.

The most straightforward way to generate ideas for a problem that you want to solve is to look at how others have solved very similar problems. As an example, let's think about how to get citizens to vote in their national elections. A striking feature of US presidential elections is that many eligible voters simply don't vote. In the 2008 presidential election between Barack Obama and John McCain, to give a fairly typical example, less than 60% of eligible voters cast a ballot and more than half of all potential voters between the ages of 18 and 24 didn't vote. In the 2020 election between Donald Trump and Joe Biden voter participation reached record high levels; about 2/3 of eligible voters cast a ballot.

For the sake of this discussion, let's assume that having more people vote is a good thing. Our problem then is getting more people to vote in elections. Instead of drilling into the complex history and legal issues around voting in the USA, we can look at how other countries approach this problem. One country that offers a striking contrast to the USA is Australia, where voter turnout for national elections is routinely higher than 95%. One positive side effect of this high level of participation is that political parties need to appeal to a broad cross-section of voters to be elected. How do Australians manage this? First, they make voting legally compulsory. That's right, it is against the law to not vote and citizens can, at least in principle, be fined for not voting. Second, the structure of Australian elections makes voting relatively easy. Australian elections take place on Saturday, a day when most people don't have to go to work. For historical reasons, US presidential elections are held on Tuesdays, meaning that many people have to take time off work if they want to vote in person.

High voter turnout doesn't mean that Australia is a political paradise. Between 2010 and 2018, a period in which the USA had two presidents, Australia had five different Prime Ministers, including one in 2013 who lasted just 83 days. It is hard to develop a long-term strategy as a government if you are in power for less than three months. But for someone interested in having more people vote in the USA, looking at other democratic countries like Australia that have much higher voter participation rates can be an energizing source of ideas and possibilities.

Thinking about voter turnout by looking at voter turnout in other countries is a simple "like-for-like" case of generating ideas. With practice, you can learn to take ideas from one place and apply them in what at first glance are completely different settings. We saw an example of this strategy when we explored analogies between movies and technical talks. There are many differences between a multi-million dollar Hollywood blockbuster and a technical talk at a conference, but thinking about similarities between the two helped us to come up with several ways to potentially make technical talks better. To apply this approach to generating ideas, it is important to articulate the key concept that allowed a problem in the other context to be solved as simply as possible. When thinking about movies, for example, we identified story telling as one element that engages audiences in a good movie. Without needing to look in great detail at the many ways a story can be told effectively, the simple idea of "telling a story" can be a useful way to make a technical presentation more effective.

As you read the technical literature and history of your field, make a habit of trying to list the key concepts people have used to creatively solve problems. These concepts will be a valuable resource as you ask *Where Else Would That Work*? Let's consider the widespread challenge of excluding the effects of impurities or background signals from experimental data by comparing two very different physics experiments. In the 1970s, physicists were interested in counting the number of neutrinos emitted by the sun. This was no simple task, since the neutrino is an elusive elementary particle that is mainly characterized by its lack of interaction with any other kind of matter; enormous numbers of neutrinos pass through the earth every second. The physicist Raymond Davis realized that if a neutrino interacted with a chlorine atom in a common dry-cleaning fluid, then it would generate an atom of argon. Unfortunately, cosmic rays from space could also have the same effect, and any signal from cosmic rays would overwhelm what was expected to be a very weak signal from neutrinos. Davis' audacious solution was to take more than 300,000 liters of dry-cleaning fluid and place it deep underground so that cosmic rays would be absorbed by material above the experiment. This was the aim of what become known as the Homestake experiment, which took place 1500 meters underground in South Dakota in the Homestake gold mine from 1970 until 1994. The Homestake experiment, which took place deep underground and involved detectors capable of measuring a few tens of argon atoms, revolutionized our understanding of neutrinos and led to Davis sharing the 2002 Nobel Prize in Physics.

Raymond Davis used what we might call the Homestake strategy of removing background signals by going to extreme lengths to create a pristine environment. A similar approach was pursued at roughly the same time by a separate community of physicists interested in understanding the properties of catalytic metal surfaces. Metals such as platinum are important as catalysts in the catalytic convertors in car engines and in large-scale industrial processes. In their experiments, these groups would carefully prepare samples in elaborate equipment that produced vacuum pressures 10^{11}–10^{12} times smaller than normal atmospheric pressure. In the 1990s, a variety of experiments and calculations had given inconsistent results about the properties of platinum surfaces. In an incisive set of experiments in 1998, Thomas Michely and coworkers showed that this inconsistency came from the presence of miniscule amounts of carbon monoxide. Instead of trying to further exclude carbon monoxide from their experiments, their key insight came from deliberately *increasing* the impurity concentration and noting its effects. That is, they used an anti-Homestake strategy.

It is almost certain that the details of counting neutrinos and the properties of platinum surfaces are irrelevant to your own research problems. The value in looking at these two examples is that they illustrate two general strategies to tackle a broad class of problems. If the effect you are studying can potentially be confounded by impurities or similar hard to detect influences, try framing these issues through a Homestake or anti-Homestake lens. Asking *Where Else Would That Work?* with these two physics experiments will help you confront the difficult question of what confounding effects might be at in play in your work.

When you try to, in Picasso's words, steal good ideas from previous work it helps to have many examples to draw upon. This observation is closely related to the bad habit we looked at in Chapter 3 of ignoring anything that isn't immediately relevant to your work. I hope that you will begin to more deeply enjoy the history of your field and successes in other fields as sources of great ideas. To close this section, we will look briefly at another scientific success that you can use as an exercise in finding ideas for your own work.

In the 1840s, a young Louis Pasteur had just finished his doctoral studies. Chemists at that time were fascinated by chemicals that were optically active, meaning that if they were dissolved in a liquid like water the resulting mixture would rotate polarized light that passed through the liquid. Tartaric acid, derived from winemaking, was one such chemical. Pasteur worked with a very similar chemical called paratartaric acid, which was not optically active.

He found that if he allowed paratartaric acid to form crystals and looked at them carefully under a microscope, there were two crystal shapes that weren't quite identical. In fact, the two crystal shapes were mirror images of each other. Pasteur painstakingly separated the two sets of crystals by hand. Dissolving one set of crystals gave a liquid that was optically active, and dissolving the other set of crystals gave a liquid with the *opposite* optical activity. Pasteur had shown that paratartaric acid was a mixture of two different chemicals whose optical properties somehow exactly balanced each other. This discovery, which was made long before chemists knew that chemicals were made of atoms bonded together to form molecules, ultimately led to the discovery that carbon atoms typically bond to four other atoms.

A survey of members of the American Chemical Society once named Pasteur's experiment with paratartaric acid the most beautiful experiment in the history of chemistry. Joseph Gal has suggested that Pasteur's insight was enabled by his artistic training, which led him to observe closely and draw what he saw. My challenge to you is to think about the approach of careful observation in Pasteur's experiment and ask yourself *Where Else Would That Work*?

Would Flipping It Help?

Soccer, the world's most popular sport, is often called "the beautiful game". The highest levels of the sport, in addition to astonishing athletic skill, can feature eccentric haircuts and dramatic acting by players trying to convince the referee they have been fouled. In the finals of major competitions like the World Cup, it is not uncommon for the score to be tied after a full game's play plus extra time. In this situation, the game is decided by penalty kicks with a series of five players from each team taking one-on-one shots at the opposing goalkeeper.

The order of penalty kicks in a penalty shootout is decided randomly by a coin toss. The team that wins the toss, team A, goes first and alternates penalty kicks with the other team. This means that the ten penalty kicks follow the pattern

A B A B A B A B A B

There is a problem with this arrangement. Some sports statisticians and many soccer players believe that going first in a shootout confers a significant advantage. If so, it seems to be a shame that something as momentous as the World Cup final could be influenced by something as random as a coin toss. We will ignore the question of whether this effect is real or not and instead look for a solution to this problem.

Penalties are taken one at a time, so someone has to go first. And having the other team go next seems natural, so the pattern above will start as A B. But what about the next two kicks? If Team A has some advantage from going first, perhaps this can be counteracted by letting Team B continue with their kicks. If we flip the order of pairs of kicks from the traditional sequence, we get

A B B A A B B A A B

Having seen an example of having non-alternating orders in this sequence, you can imagine other possible orderings. Can you generate an example that seems fair in which team A goes first and last? The alternate ordering above has actually been tried in a small number of low-level tournaments, but the power of tradition in sport is strong, so it is unlikely that you will see it in a future World Cup.

This analysis of penalty kicks is an example of Nalebuff and Ayres' fourth question for generating great ideas: *Would Flipping It Help?* To apply this question, find ways to turn statements upside down, back to front or inside out. This often leads to unhelpful nonsense, but surprisingly often, it can help generate new ideas. If you work at a food company, turning "Tacos make a great evening meal" into "Tacos make a terrible evening meal" is not too helpful. In contrast, another way of flipping the statement gives "Tacos make a great morning meal", which might actually prompt the company to think of new markets.

Let's look at two examples from the world of transportation. In many cities, tolls associated with bridges or tunnels are important sources of revenue for local governments. If even a fraction of the traffic has to stop to physically pay a toll, the resulting toll booths require space and, more importantly, can create a traffic choke point. Imagine that your objective is to collect tolls from drivers and to limit the resulting traffic jams. The usual approach is to charge a toll on people going from A to B on the road and also on people going from B to A. One way to flip this situation is to not charge anyone tolls, but that isn't helpful. Another way to flip things would be to charge tolls for going from A to B but not for the reverse journey. Since most local traffic goes one way in the morning and the opposite way in the evening, this flipped approach generates almost all of the tolls with approximately half the traffic interruption. This solution makes so much sense that it is quite common – you may be able to think of a toll road like this in a major city you are familiar with.

A second transportation example is probably familiar to you from going to the airport. Many airlines, at least in the USA, use a complex hierarchical boarding system ("We will begin boarding First Class and Lifetime Titanium Members, followed by Iridium Class Guests and Golden Ticket Pass Holders, then continue with Comfort Select seating before moving on to Boarding Groups 1, 1a, 2, 3 and Economy Survival Class"). An inevitable outcome of this system is that boarding a plane takes a really long time, often 30 minutes or more. If you have ever shuffled along at the back of the line for this process, you have undoubtedly wondered if it was possible to get everyone on a plane more quickly.

Can we get people boarded on a plane quickly? Let's flip the question. Can we get people on plane slowly? Probably, but that isn't going to help. Can we get people off a plane quickly? The answer, of course, is yes. Airlines are required to be able to get everyone off a plane extremely quickly in the unlikely event of an emergency. Can ideas from that process be adapted to boarding the plane? One of the keys to getting people off a plane in an emergency is to have people leaving from multiple exits. Large aircraft have, minimally, doors at the front, doors at the back and emergency exits in the middle of the cabin.

Have you seen how to speed up plane boarding? Most airlines use only a single forward door during boarding, meaning passengers have to move in single file through the narrow aisle in the plane. This speed can be at least doubled by also using a door at the back of the plane. Just one major US airline, Alaska Airlines, uses this approach, although it is more widespread in other countries. Boarding through the back door means that passengers have to walk outside on the tarmac, but unless the weather is truly awful, many people seem happy to do this in order to get to their plane seat faster. And the Lifetime Titanium Members and their bulging carry-on bags can still enter through the forward door.

The key observation from the examples of toll roads and plane boarding is that flipping the problem statement made generating actionable solutions almost obvious. Arguably, the idea of peeling a banana from the other end could be "discovered" using the flipping strategy without needing any observation of monkeys or similar inspiration. We saw an application of this idea to scientific research in the previous section where we discussed the Homestake strategy of going to extraordinary lengths to exclude impurities or background signals from an experiments. Flipping this idea immediately suggests the anti-Homestake approach of deliberately adding impurities to understand what will happen. Simply using the name "anti-Homestake" is a giveaway that this is an idea that can come from flipping something. Try applying this simple linguistic trick to some of the methods you use in your work as a way to generate new ideas for your work.

One of the pleasing things about ideas that come from the flipping approach is that they often seem obvious in retrospect. Several years ago I was part of a large research center that wrote many papers and spent millions of dollars in research funding studying degradation of materials. We specialized in exposing materials to acid gases and analyzing the negative effects these had on various technologically relevant materials. A creative PhD student in the center, Krishna Jayachandrababu, asked a deceptively simple flipped question: could we use the acid gases to make a material better instead of worse? It turned out the answer was yes.

Krishna developed a method in which acid gases were used to partially "deconstruct" a material that could then be "reconstructed" to be different and in some ways better than the original. This idea didn't emerge in a single flash of insight – it took months of painstaking trial-and-error work. The key idea that motivated this work, however, came from flipping a problem statement that many others had accepted without challenge and seeing where that led.

What Would Gandhi Do?

We began this chapter by asking *What would Croesus do?*, which motivates us to look at a problem as though we had no resource constraints. If we flip this question, we instead look for solutions that use almost no resources. In other words, *What would Gandhi do?* This question forms the basis of an active field called frugal science that aims to replace expensive scientific equipment with extremely cheap devices made from commodity components that do most if not all of the job of the original. Researchers in frugal science don't aim to reduce equipment costs by 10% or a factor of two; they want to make instruments that are almost free and can be used by anyone anywhere in the world.

One example of a success in frugal science comes from my Georgia Tech colleague Saad Bhamla, whose research group developed a low-cost device called the ElectroPen. This gadget is used for electroporation, a technique that uses electrical fields to cells to deliver molecules such as DNA or vaccines. Although they are used in biology labs around the world, electroporation equipment often costs thousands of dollars and, for obvious reasons, require electricity. The Georgia Tech group developed a device from a hand-held stove lighter (channeling another one of this chapter's questions, *Where else would that work?*) that requires no electricity (!) and can ultimately be made for $0.23.

Chapter Summary

In this chapter, we have looked at systematic ways you can use to regularly generate outrageously good ideas. To illustrate tools for tackling this task, we have looked at a menagerie of good ideas from others, from ultra-cheap lab instruments to voter turnout to giving non-boring talks. As you look over the chapter, remember that the aim here is not to list examples of good ideas but to look at systematic strategies that you can borrow to come up with ideas of your own.

Being good at generating ideas isn't a mystical ability; using a series of prompting questions can help you look at problems in new ways. We have focused on four prompting questions from Barry Nalebuff and Ian Ayres. Next time you have a problem to solve, work through these questions:

- *What Would Croesus Do?* (and its flipped companion question *What Would Gandhi Do?*)
- *Can You Tell Me Where It Hurts?*
- *Where Else Would That Work?*
- *Would Flipping It Help?*

Further Resources for Chapter 4

The earliest discussion of banana peeling methods I know of is a 2002 column in the online magazine *Slate* by Stephen E. Landsburg titled "The Great Banana Revolution: Should You Peel Bananas from the Bottom Up?".

The four main strategies described in this chapter come from *Why Not? How to Use Everyday Ingenuity to Solve Problems Big and Small*, Barry Nalebuff and Ian Ayres (Harvard Business Review Press, 2003).

More information about Stuart and Cedar Anderson's flow hive can be found at www.honeyflow.com.

To learn more about the "customer discovery" focus of the lean start up approach, *The Lean Startup: How Today's Entrepreneurs Use Continuous Innovation to Create Radically Successful Businesses*, Eric Ries (Currency, 2011) is a great starting point.

For more details on the influence of minute amounts of carbon monoxide on platinum surfaces, see How Sensitive is Epitaxial Growth to Adsorbates? M. Kalff, G. Comsa and T. Michely, *Physical Review Letters*, 81 (1998) 1255.

The connection between Louis Pasteur's artistic training and his experiments with crystals of paratartaric acid are explored in Pasteur and the Art of Chirality, Joseph Gal, *Nature Chemistry*, 9 (2017) 604. Although Pasteur's experiment is often described today as a heroic leap forward in chemistry, the truth is almost certainly more complicated. Phillip Ball describes the history of these experiments and the many elements of luck that led them to work in his book *Elegant Solutions: Ten Beautiful Experiments in Chemistry* (Royal Society of Chemistry, 2005).

The first mover advantage in penalty kicks and potential alternate orderings for taking penalties in soccer were discussed by Umair Irfan in *Why soccer teams that go first in shootouts usually win* (https://www.vox.com/2018/7/11/17537886/world-cup-2018-penalty-shootouts-kicks).

Krishna Jayachandrababu's material "healing" process is described in Recovery of Acid-Gas-Degraded Zeolitic Imidazolate Frameworks by Solvent-Assisted Crystal Redemption (SACRed), K. C. Jayachandrababu, S. Bhattacharyya, Y. Chiang, D. S. Sholl and S. Nair, *ACS Applied Materials and Interfaces*, 9 (2017) 34597.

The ElectroPen is described in ElectroPen: An Ultra-Low-Cost, Electricity-Free, Portable Electroporator, G. Byagathvalli, S. Sinha, Y. Zhang, M. Styczynski, J. Standeven and M. S. Bhamla, *PLoS Biology*, 18 (2020) e3000589. The ElectroPen isn't the only frugal science success story from Saad Bhamla's lab – they have also developed a hearing aid that costs $1 and a hand-powered centrifuge for molecular biology that they have used in high-school classrooms and the Amazon rainforest.

5

Writing for Fun, Profit and Career Advancement

Long-term success in a career focused on research is not possible without being able to express yourself clearly in writing. Completing a PhD requires writing a thesis that communicates the outcomes of years of hard work and thinking. Writing papers that impress reviewers, journal editors and ultimately your professional peers is vital for sharing your work with the technical community. Writing proposals that convey the excitement of your ideas is crucial for sustaining any research group and allowing you to explore your next set of groundbreaking ideas. In a job interview, no one would volunteer that they were overflowing with good ideas but were terrible at technical writing. Despite the critical nature of this aspect of a career in research, many researchers view writing with similar enthusiasm as a trip to the dentist. In this chapter, we will look at strategies that will allow you to improve the quality and efficiency of your writing, and, dare I say it, even make writing something that you will look forward to.

This chapter addresses writing that aims to deliver technical information in a clear and compelling way. Professional poets or authors of literary novels spend huge periods of time making word choices or adopting sentence structures that will perfectly convey emotional nuance or otherwise deliver their artistic vision. These pursuits play an important role in human well-being, but they are irrelevant for our purposes here. When writing about your research, you are aiming to clearly articulate your technical findings to an audience who are short on time and scientifically critical and who come from all over the world. In this context, clarity is highly valued, but poetry may be counterproductive.

In the first part of this chapter, we look at some remarkably simple habits that can transform your effectiveness as a writer. Many of the ideas in this section are adapted from a wonderful book by Paul Silvia titled simply *How to Write a Lot: A Practical Guide to Productive Academic Writing*. Professor Silvia is a successful academic psychologist at the University of North Carolina who has published two other technical books and over 100 papers. He has a knack for incisive writing. One of his highly cited papers with the intriguing title "What is interesting? Exploring the appraisal structure of interest" appeared in the journal *Emotion* in 2005. I like to imagine that in addition to their opinion on the correctness and value of the research, the reviewers for this journal are asked "How did the manuscript make you feel?".

After describing habits for effective writing, I turn in the second part of the chapter to how to use your time when writing. To preface this topic, we must define what activities count as writing. Most people, if asked to define what writing is, would think of the specific actions associated with typing words into a document or editing a draft piece of text. This is certainly correct, but the act of writing for a researcher also encompasses many other actions. I like Paul Silvia's definition: "Any action that is instrumental in completing a writing project counts as writing". Analyzing data to explore research hypotheses or preparing figures definitely counts as writing. Following your favorite celebrity influencer on social media or chatting about your sports team's weekend performance is definitely not. Reading key background literature to understand the latest ideas in your field? Writing. Tinkering with the slide template in the slide deck for an upcoming talk? Not writing.

By focusing on a "writing project", Silvia's definition brings out an underappreciated aspect of professional writing. Some researchers think of writing as something that is done as a separate activity after their research is done. I think this is a serious mistake. If you agree that reporting your research as a paper, thesis, or in some equivalent form is a fundamental requirement for the research to be finished, then you should consider your work to be a "writing project" from its earliest stages. Drafting an introduction to your paper *before* your work starts is a powerful tool to focus your hypotheses and aims. The details of this introduction are likely to change as your work and understanding progresses – this is a healthy process. Many effective researchers use writing in an iterative way in developing their ideas, with initial drafts of figures and text driving additional thinking, experiments and analyses. The process of working through multiple drafts of a written document in which real changes in ideas and content take place between drafts is a sign of thoughtful and creative research. This goal cannot be reached if writing is viewed as an irritating but necessary "add-on" activity that is performed after all the real work has already been done.

As we move through this chapter, we will meet several prolific writers who can offer inspiration to all of us that make writing part of our professional lives. The first example is Anthony Trollope, a British novelist who lived from 1815 to 1882. Photographs of Trollope show a bespectacled figure with the kind of expansive facial hair that was favored by Victorian gentlemen. Trollope wrote more than 60 books, more than many people read in a lifetime. His novels were widely read while he was alive and are still highly regarded.[1] One of the amazing features of Trollope's life is that he was not

---

[1] Because the copyright on Trollope's books has expired, they are available very cheaply. One collection of 47 novels plus other assorted works is available electronically for just a few dollars. If you are looking for some reading material for an upcoming long haul flight or extended vacation, it would be hard to find better value for money.

a full-time writer. For much of his life, he worked at the British Post Office. This was not a low-level position delivering mail. Among other things, Trollope reorganized the structure of the postal service for all of southern England and introduced the concept of the pillar box to collect mail. He is famous today for the regularity of his writing habits; he rose early each day and wrote for several hours before heading to his "real job". It is reported that if he finished a book in the middle of one of these writing sessions, he would immediately start working on his next book instead of waiting for the next day. Trollope's thousands of daily writing sessions may seem a touch monomaniacal, but they are powerful reminder of what can be achieved by steady, sustained work.

A more modern example of a prolific writer who did his writing "on the side" is Jim Lehrer. Lehrer, who died in 2020, was the co-anchor of an hour-long nightly news program on the American public TV network PBS from 1975 to 2011. He also moderated 12 debates between presidential candidates in US elections between 1988 and 2012, earning him the nickname "The Dean of Moderators". It would be easy to imagine that the rigid deadlines of broadcast journalism would not leave Lehrer with much time for other pursuits, but that was not the case. Lehrer wrote regularly on "nights and weekend on trains, planes and sometimes in the office", producing more the 20 novels, three memoirs, and multiple plays.

The literary output of Anthony Trollope and Jim Lehrer would be impressive if their entire lives had been devoted to writing. The fact that both of them had demanding full-time careers that were entirely independent of their writing makes their accomplishments even more striking. When confronted with examples like this, most people think "I could never do that". Before you read further, take a few moments to list some of the reasons that you find technical writing difficult and why Trollope and Lehrer seem almost superhuman. Next, we will confront some of the most common challenges we all face in attempting to write.

Specious Barriers to Writing and the Secret of Successful Writing

The idea that effective writing is critical to career success is uncontroversial. So why is writing viewed by most researchers with a mix of guilt, frustration and loathing? Paul Silvia analyzed this question by articulating a series of reasons that researchers at all career stages give to justify their lack of progress in writing. He calls these excuses "specious barriers" because while they sound superficially plausible, more thoughtful consideration shows that they are rarely well grounded. Let's look at these specious barriers one by one.

Specious Barrier #1: I Can't Find Time to Write (a.k.a. I Could Write Much More If I Just Had Large Chunks of Time)

Academic researchers love to complain (or brag) about how busy they are. When was the last time you heard a colleague say they were happy with the balance they had achieved between all the activities going on in their life? Telling yourself or others that you just don't have time to do the writing you need to do can almost seem like a badge of honor. Much of this perceived busyness is real – many people want to or need to attend to family commitments and similar important aspects of a fulfilling life in addition to all the aspects of work needed to maintain a successful career. At the same time, these kinds of time pressures also exist for the people you can think of that produce large numbers of papers, research proposals and similar writing products every year.

Silvia points out that the wording of this barrier implies that our time is somehow out of our control and that we are forced against our will into a situation where we don't have time to write. Under ordinary circumstances, however, most of us have other time-consuming activities that we attend to without fail.[2] If you are an Assistant Professor or a Teaching Assistant, you presumably don't tell people that you couldn't teach your class this week because you couldn't find time. When your lecture is scheduled from 10 to 11 am, you plan ahead, block the time on your calendar and show up to teach the class. If you have an appointment with your dentist, you make time in your schedule and arrive at their office on time so you can recline in their chair and attempt to make small talk while they poke tools in your mouth. These two examples illustrate the truism that time isn't something we "find", it is something we control and manage through our decisions.

Looking at time management in this way leads to an almost inescapable conclusion: if you agree that writing is a vital part of your career success, then you should schedule specific times to write and stick to them. Silvia put this eloquently – *"Prolific writers make a schedule and stick to it. It's that simple"*. He recommends starting by allocating four hours per week and treating these appointments with the same respect and protectiveness that you already give to teaching, going to the dentist and so on.

For the past seven years, I have been the Department Head for a large academic department while also working with a productive research group of hard working PhD students and postdocs. For at least the last five years,

---

[2] Many people go through seasons in life when they have truly overwhelming demands on their time – giving birth and caring for a newborn or dealing with a life-threatening illness for themselves or a loved one, for example. If you know someone in a situation like this, don't complain to them about how busy you are. Instead, ask them if there is some way you can help them.

I have also followed Silvia's advice – my weekly calendar shows blocks of time marked "Research" from 7:00 to 9:30 in the morning on Tuesday and Thursday. I view these blocks as uninterrupted time when I work on writing the manuscripts my group and I are drafting and the research proposals necessary to keep my research group funded. Scheduling this time has value in multiple ways. Externally, the existence of these periods on my calendar gives me a way to graciously say no to people who want to schedule meetings during those times. (The number of people who want to meet at seven in the morning is small but nonzero, but I can always talk with them on Monday, Wednesday or Friday.) Much more importantly, however, these periods give me mental permission to ignore my email in-box and all similar non-writing tasks for a short period. The strategy of scheduling writing time and sticking to it has been instrumental in helping me to maintain a steady stream of research publications while also working with my faculty and staff colleagues to keep our academic department running smoothly.

The "schedule it and stick to it" approach will not only improve your writing productivity but also dramatically reduce the existential angst many researchers feel about their writing. For years, I carried around a constant low level of guilt because students in my research group had written initial drafts of papers that then languished for weeks or months on my computer. I knew I needed to work on these pieces of writing, but as the specious barrier we are discussing says, I would tell myself that I just couldn't find the large chunks of time I would need to do so. Since training myself to schedule writing on my calendar, I still worry from time to time about editing a student's thesis chapter or manuscript promptly, but now I can remind myself that this task can be tackled at a known time within a few days. It may sound over the top to say that a strategy like this can be life changing, but at least for me, this is literally true. If you take away one thing from the chapter, I urge you to try Silvia's strategy for yourself and see how it works. As Silvia wrote "At first, allot a mere 4 hours per week. After you see the astronomical increase in your output you can always add more time".

The "large chunks of time" portion of this specious barrier does contain a kernel of truth. At the start of any writing session, it will take at least a few minutes to focus your thoughts. This means that breaking your writing into a large number of short periods will be inefficient. Effective writing requires sustained concentration, so unless you have truly heroic levels of self-discipline, the efficiency of your work will go down after a couple of hours. For this reason, splitting Silvia's recommended four hours into two sessions of two hours each seems about right. (If the thought of concentrating on a single task for two hours sounds impossible to you, go back and reread Chapter 2 on the power of deep work.)

While scheduling regular times to write can reap enormous dividends, other variations of the "schedule it and stick to it" strategy can also be effective. My job involves a lot of travel. One way to look at this is that I get to spend many hours every year sitting in airport lounges and airline seats.

A more positive way to think about this situation is that my calendar frequently features hours-long chunks of time when I am cut off from the outside world. Several years ago, I decided to use this time for writing. By cultivating this habit, I was able to start and complete a novel.[3] In this case, my "schedule" for writing was initiated each time I arrived at the airport. The point of this anecdote is not that my situation will apply exactly to your life. Rather, it illustrates how thinking carefully about your schedule and, more importantly, training yourself to write at specific times or in specific settings can do wonders in dispelling the specious barrier that you don't have time to write.

Prolific writers make a schedule and stick to it. It's that simple.

Specious Barrier #2: I Need to Read a Few More Articles

In his science-fiction comedy classic "The Hitchhiker's Guide to the Galaxy", Douglas Adams wrote "Space is big. You just won't believe how vastly, hugely mind-bogglingly big it is. I mean, you might think it's a long way down the road to the chemist's, but that's just peanuts to space". This is also a good description of the scientific literature. One journal my research group consults regularly is the *Journal of Physical Chemistry C*. In 2018, this journal published more than 29,000 pages of content. In case that isn't enough reading material, there are also the *Journal of Physical Chemistry A*, the *Journal of Physical Chemistry B*, and the *Journal of Physical Chemistry Letters*, not to mention the *Journal of Chemical Physics*, *Chemical Physics Letters*, *ChemPhysChem*, and *Physical Chemistry Chemical Physics*, all of which are respected scholarly journals. Without even going beyond journals whose titles are just variations of the words "chemistry" and "physical", there were well in excess of 100,000 pages of new journal articles published in just a single year. This is not a phenomenon that is specific to chemical research; every area of research has a set of legitimate journals that contain a vast and continually expanding collection of information.

A truism of scientific research is that "an hour in the library can save you weeks in the lab". Developing the aims and approach for any new project must involve carefully reviewing and understanding existing work. Failing to do so risks spending your time reinventing the wheel (at best) or enraging peer researchers who feel that their work is being ignored. It is easy, however, for time spent on this kind of background work to become a procrastination strategy. It is sometimes easy to detect when your reading has become a

---

[3] I make no claims that the resulting book is a literary masterpiece, but am happy to report that several people who are not related to me have told me that they enjoyed reading it.

diversion from your writing. If your Internet searches have led you to pages of the "12 Greatest Low Budget Movies of the 1990s" variety, then progress towards your writing goals has almost certainly stalled. What has become known as Zuckerberg's Law can be a useful heuristic: higher rates of Internet use correlate with shallower thinking. The enormous size of the research literature in every field, however, means that there are always "a few more articles" to read that could actually be related to your work. A key question is not whether there are more relevant articles out there (if you take a relatively broad view of what your research encompasses, the answer to this is always yes) but when to stop.

In deciding whether doing background reading has become procrastination, it is helpful to remember Silvia's definition of writing: "Any action that is instrumental in completing a writing project counts as writing". Since you are presumably interested in doing high-quality research, sharpening your understanding of the details of the topic of your writing project will be instrumental in completing that project in a satisfactory way. Spending time thinking deeply about papers or other sources that are directly related to your work is always time well spent. Most kinds of scientific and technical writing don't face the same kinds of strict deadlines faced by, say, newspaper journalists.[4] It is important to leave open the possibility that something you learn during a writing project will require significant rethinking of your work or additional experiments or analysis. It is not uncommon to think that a manuscript is "almost finished" only to realize that significant additional effort is required to have a complete product. When this happens, spending time reading relevant literature in a focused way can be vital. If, however, you are reading superficially or gathering reading material indiscriminately, it is likely that you are using these actions to avoid the hard work needed to truly move your writing forward.

The idea of reading "a few more articles" is one of several procrastination strategies that are tempting justifications for lack of progress in writing. Fiddling with the formatting of a document, adjusting the color scheme and appearance of figures, and perfecting a manuscript's bibliography are all examples of activities that contribute to the final quality of your writing but can easily become time-consuming diversions. One way to assess whether you are falling into this trap is to imagine describing your division of time to someone else – "this week I spent five hours writing and of that time I spent two hours making a key figures for my manuscript and the rest of the time writing a first draft of my Methods section". If sharing an honest accounting of your time with someone else would make you uncomfortable then it might be worth carefully assessing how that time is being spent.

---

[4] An exception is writing research proposals, which often have well defined and unforgiving deadlines. These deadlines, however, are typically known months in advance.

Specious Barrier #3: I Need a Nicer Chair/ New Computer/Stronger Coffee

Stephen King is one of the world's most successful writers. He has written more than 60 books and 200 short stories, and over 350 million of his books have been sold. King has a pithy description of the physical and mental environment that is needed for effective writing: "The space...really only needs one thing: a door you are willing to shut". He recommends writing in a room with no windows or the blinds shut with a desk in the corner! Generations of writers did their successful writing by hand on paper or using manual typewriters in drafty rooms with poor heating or cooling. It is unlikely that the specific computer you are using or the chair you are sitting in is really the key barrier stopping you from writing effectively.

Years ago I taught a graduate-level engineering course that met for 90 minutes in the middle of the day. In the middle of the semester, I surveyed the students to ask what was working well in the class and what could be improved. One memorable respondent wrote "I am always hungry when I come to this class because it is at lunchtime". Sitting in a challenging lecture with a rumbling stomach is certainly not ideal, but the student's response is a good example of a specious barrier. If being hungry is stopping you from concentrating on a class you attend, then have a filling snack before class. If not having coffee is stopping you from writing, make some coffee, then move past your specious barrier and do some writing.

Specious Barrier #4: I Am Waiting Until I Feel Like It (a.k.a. I Write Best When I'm Feeling Inspired)

A popular image of a great artist involves a free spirit who creates inspiring works when they are filled by some kind of mystical creative energy. This rarefied state of mind seems to be antithetical to the constrained idea of scheduling time for writing. Since scientific research is meant to be a creative endeavor, this image can lead to the idea that researchers should "wait for inspiration to strike" before working on writing. Setting aside the question of whether this stereotype has any validity for creative artists, it is important to realize that viewing technical writing as an other-worldly creative experience is wrongheaded. Borrowing an analogy from Silvia, a scientific writer shouldn't think of themselves as Pablo Picasso painting "The Old Guitarist"; they should think of themselves of the guy from "Smith and Sons Painting" who you have hired to repaint the front of your house on Thursday.[5] The homeowner expects Smith (or his son) to paint the house on time and with a minimum of fuss, not to wait until they feel inspired to work. It can be helpful to view writing in the same way. Sinclair Lewis, who won the Nobel Prize in Literature, described an influential instructor who pithily summed up the task of a writer by saying "The art of writing is the art of applying the seat of the pants to the seat of the chair".

Amusing analogies are all very well, but the specious nature of this barrier is illustrated in a more compelling way with actual data. Robert Boice collected data for this purpose as part of his book *Professors as writers: A self-help guide to productive writing*. He studied a sample of 27 university faculty who identified themselves as struggling with writing who were randomly assigned one of three writing methods for a ten-week period. The "abstinence" group were told they could do no non-emergency writing, the "spontaneous" group were asked to schedule 50 writing sessions but to only write during these sessions if they felt inspired, and the "scheduled" group also scheduled 50 writing sessions but were told to write regardless of how they felt. Subjects in the latter group had to send a donation to an organization they despised for each writing session they skipped.[6] Not surprisingly, the scheduled writers wrote a lot more than the other groups: on average they wrote 3.2 pages per day, compared to 0.9 pages per day for the spontaneous writers and 0.2 pages per day for the abstinent group. One of the subjects from the "scheduled" group said "It really isn't what I thought it would be…. What I really like…is how easy it is to start writing. No struggle. I look forward to it…Sometimes I'm tempted to start sooner". Does that describe how you feel about writing?

As a skeptical researcher, you are already thinking to yourself "Sure, those scheduled people wrote a lot, but was it any good?". Boice also had his subjects record how often they had what they considered a creative idea in their work. This seems like a more meaningful measurement of research progress than just counting the number of words or pages someone has written. The abstinent writers reported having a creative idea on average every five days, and the spontaneous writers had a creative idea every two days. The scheduled group, however, reported creative ideas at an average rate of one every day. Stop and think about that for a while: simply by scheduling their writing time, Boice found that he could double the frequency with which his academic subjects had creative ideas. Are you still going to "wait until you feel like it" to work on your writing?

---

[5] In fairness to Picasso, he produced more than 10,000 artworks in his lifetime, so he was either filled with creative energy on a very regular basis or he got on with his work without waiting until he felt like it.

[6] This was the same idea that was turned into an online business venture by Ian Ayres, the Yale law professor we met in Chapter 4.

There is an interesting postscript to Boice's data. Remember the "scheduled" writers who had to send money to a group they didn't like every time they missed a writing session? Boice defined a failed writing session as a day that didn't produce three pages of text. It hardly seems coincidental that the average amount of text produced per day from this group was 3.2 pages. This suggests that in defining your own writing schedule, it is worth experimenting with the measures you use to define success.

Even if you buy into the idea of scheduling your writing, there will be times when you just feel completely stuck. The cursor on your screen will blink maddeningly at you, but you seem unable to string a sentence together, let alone move your writing ahead and generate creative ideas. Faced with this, many people find it useful to write a "terrible" draft that just aims to get something written down, ignoring any concern for spelling, punctuation, grammar or logic. Try doing this for five minutes, then walk around the room to stretch your legs and choose one or two elements of the word salad you have created to edit or modify. You don't have to ever show your terrible draft to anyone or make any kind of judgment about whether it is "good" or not. Your draft is useful simply because you have put words together and written something down. If even beginning a terrible draft feels difficult, imagine that you are describing what you are trying to write to a supportive and patient colleague or friend. Even better, use a recording app on your phone and record yourself as you actually talk for a minute or two, then transcribe what you said in writing. Again, don't worry about trying to sound eloquent or clever; just say (and then write) whatever comes to mind.

Moving Beyond Specious Barriers

By framing a discussion of writing in terms of specious barriers, Paul Silvia has done a service to researchers of all kinds. The list above is intended to make you state potential excuses you might give for not writing ("I don't have large chunks of time" or "I am waiting until I feel like it") and think objectively about whether those excuses are justified. The main point of this discussion is simple and worth repeating in Silvia's words: "*Prolific writers make a schedule and stick to it. It's that simple*". The benefits of this scheduled approach to writing are plentiful. You will write more. You will have more creative ideas. Your mental health will improve because your guilt about writing projects that aren't progressing will dramatically diminish. If these outcomes aren't appealing to you, that is your choice. But if they sound attractive, try something like the ten-week experiment of scheduled writing that Robert Boice conducted and see for yourself how it works.

As you make a plan to schedule your writing time (and protect that time from email, meetings and all the other distractions of the modern workplace), remember that how you organize this time is just as important as the total amount of time you use. A good example in this regard is the widely acclaimed science-fiction author Neal Stephenson. Stephenson has written 16 books to date, including *Cryptonomicon*, a thriller that combines World War II code breaking, the history of computation and anticipated cryptocurrencies years before Bitcoin was invented, *REAMDE*, a sprawling adventure involving the developers of a multiplayer computer game, Chinese hackers, Russian mobsters and Islamic terrorists, and *Seveneves*, in which the moon explodes in the opening chapters and the rest of the book follows the next several thousand years of human history. As well as being mind-broadening and hugely entertaining, Stephenson's books are long: the combined paperback versions of just the three I mentioned have more than 2800 pages. On his website, Stephenson has a page explaining why he doesn't answer email from readers or use other kinds of social media. Here is what he says:

> Writing novels is hard, and requires vast, unbroken slabs of time. Four quiet hours is a resource that I can put to good use. Two slabs of time, each two hours long...are not nearly as productive as an unbroken four....The productivity equation is a non-linear one, in other words...If I organize my life in such a way that I get lots of long, consecutive, uninterrupted time-chunks, I can write novels. But as those chunks get separated and fragmented, my productivity as a novelist drops spectacularly.

For the sake of your professional colleagues, I hope your next writing product isn't a 1000-page single-spaced manuscript reminiscent of a Stephenson novel. But everything he says applies to technical writing. Technical writing is hard and requires slabs of time for any progress to be made. If you organize your life in a way where you give yourself lots of long, consecutive, uninterrupted time chunks, you can massively improve the quantity and quality of your technical writing. If you let those chunks get separated and fragmented, your productivity as a writer will drop spectacularly.

What to Do When You Write

Congratulations! You've scheduled time to write, you've mentally blocked out the other demands on your time and distractions, and you are ready for your professional productivity to soar. You sit down at the beginning of your first writing session and....what do you do exactly? In the rest of this chapter, we'll look at how you can most usefully spend your time while writing. I am not going to tell you how to write better and more grammatically correct prose; there are many sources far better equipped than me in this area. (*The Elements of Style* by Strunk and White is a classic example, with timeless advice such as "Omit needless words".[7]) Instead, we will focus on how to spend your time when writing.

Imagine for a moment that you would like to become an accomplished piano soloist. How should you spend your time? Listening carefully to concerts and recordings of other artists would certainly be useful, but you could do that for an awfully long time and still not be able to play the piano yourself. Instead, the answer is in the time-worn joke about a musician who asks a cab-driver in New York City how to get to Carnegie Hall, only to be told "Practice, practice, practice!" In the same way, spending your time reading high-quality technical writing will definitely improve your long-term writing, but it can't teach you what to do while you are writing yourself.

There are many parallels between the skill of practicing a musical instrument and the act of writing. To explore this analogy, I will use a series of ideas from a terrific book for musicians by Tom Heany called *First, Learn to Practice*. Heany argues that musicians can greatly benefit from carefully thinking about how they practice, and the same can be said about a researcher who wants to write. Below, we will explore several key points from Heany's book and how they can help you write more effectively.

Playing and Practicing are Two Different Things

Watching an accomplished musician perform a piece can be exhilarating. Watching the same musician practice, however, is likely to be frustrating or even boring. Professional musicians rarely practice by playing through long sections of music. Instead, they tend to repeat short snippets of music many times, experimenting with and perfecting tiny variations in phrasing and fingering. In a similar way, effective writers don't sit in a quiet room and write an entire book chapter or technical paper from beginning to end. Effective technical writing requires splitting the overall aim of the writing project ("I need to write my PhD thesis") into a set of many sub-goals that can be tackled in a meaningful way ("I need to list the key items to address in the discussion of my results in Chapter 3" or "I need to decide if the draft figures in Section 2 are clearly illustrating the things I want readers to notice in the data").

When you begin a writing session, start with a specific goal or a few goals. The aim of these goals is not to finish your writing, but to move your writing project forward. Your goals should be specific enough that you can easily

---

[7] The famous poet Dorothy Parker said, hopefully tongue-in-cheek, "If you have any young friends who aspire to become writers, the second-greatest favor you can do them is to present them with copies of *The Elements of Style*. The first-greatest, of course, is to shoot them now, while they're happy."

determine if you achieved them and limited enough in scope that you can usually accomplish them within a single writing session. Many people find the habit of breaking a large writing project into many small pieces to be liberating. It is far more useful to develop a series of small steps that will move your writing forward than it is to lay awake at night repeating "I have to write my thesis tomorrow" over and over.

When a musician practices a small section of complicated piece, their aim is not to capture a single perfect rendition. The aim of this kind of practice is to improve a future performance. It is useful to think about your writing sessions in this way – the aim is to move your writing project forward, not make every piece of your document absolutely final. You shouldn't iterate on a piece of writing forever, of course, but it is often surprising how quickly a draft document will converge to a complete version once you make a concerted effort to list the individual items that you need to work on and then tackle them in turn.

Don't Worry about the Hard Part

A performer preparing a piece of music for performance typically finds that there are some short sections of the music that are far more difficult than the rest of the piece. Often, the rate at which notes have to be produced increases alarmingly and the techniques needed to generate the desired effect become far more advanced. It is here, at the hard part of the piece, where one slip can bring the whole performance to a disastrous collapse, an event musicians laconically refer to as a "train wreck". Given this reality, the advice to not worry about the hard part may sound like crazy talk. Heany's idea isn't that musicians should avoid problems by just thinking happy thoughts or visualizing themselves giving a flawless performance. Instead, he is pointing out that simply worrying about something isn't useful. It is far better for a musician to consciously pinpoint the hard part of a piece and then to carefully work on that section. This kind of work often begins by moving very slowly until the exact sequence of muscles and fingerings becomes second nature. (Remember how I said that listening to a musician practice can be boring? This is why.) When this approach is followed diligently over many hours of practice, the hard part of the piece gradually becomes less frightening, then just a moderately difficult, then ultimately just another part of the overall piece. Worry has been replaced by mastery.

A large writing project, just like a piece of music, often has one or more "hard parts" that cause more anxiety than everything else. Perhaps there is some key data that doesn't follow the trends you expected. Perhaps you can't see a way to connect the aim of your work with the ideas that are considered important in your subfield. Or perhaps you have a nagging sense that the kinds of experiments you have done have already been reported somewhere else. Just like a master musician, the best approach is to consciously identify the issue that is worrying you and to identify specific steps you can take to work on the issue. Ignoring the issue or just worrying about it will not make it go away. Just like a practicing musician, this process will involve real work and can't yield instant results. You may realize that additional control experiments are needed or that you have to write a nuanced discussion that openly discusses several possible interpretations of your results without hyping a single hoped-for conclusion.

What you consider to be the "hard part" of a writing project will likely change over time. At the earliest stage of writing a research proposal, the task of coming up with a central idea is almost always the most difficult thing to do. Once this (very challenging) task has been addressed, the hard part might be developing a specific plan for a key sub-aim of your proposed work or it might be writing a compelling half-page introduction that gives credit to existing work in the field but lays the groundwork for your exciting ideas. At each stage, Heany's approach of identifying the topic of the greatest current concern and then breaking that topic into a series of specific tasks that are individually manageable is a powerful antidote to worry. Even better, by continuing with this practice, a day will come when you realize that there are no "hard parts" left in your writing project. When this happens enjoy it! There may still work to be done, but you are well on your way to finishing your project.

Be Honest

Many more people sing in the shower than sing in public. There are good reasons for this. The enclosed space in a bathroom creates echo and reverb that makes singing voices sound fuller than they really are. Just as critically, a shower performance is given for your own entertainment, not for anyone else to listen to. As you approach a tricky part of your aquatic rendition of *Stairway to Heaven*, Robert Plant's distinctive vocals can be replaced by a short air guitar solo and the music you are imagining in your head sounds just as good. This scenario highlights a key challenge for musicians seeking to practice effectively: a musician knows how their piece is meant to sound and their brain can easily "fill in the gaps" between the sounds they are actually making and the intended result.

To improve, a musician must be honest about the results they are actually producing as they would be heard by an audience. As a writer, you must consider whether your readers will understand your writing given only the information in the document you have written. It doesn't matter if you meant to include some critical background information or details about your research methods – if that information isn't in your document, then it is not available to your readers. In the paragraph above, if you don't know what *Stairway to Heaven* or air guitar is then my weak attempt at an amusing illustration merely caused confusion. (A quick Internet search for videos on either, or even better both, of these topics should at least provide a few minutes of entertainment.)

For a musician, a key approach to being honest is to periodically make an audio or video recording of a practice session and then listen critically to the result. Think about recording yourself sing a well-known song and then listening to your recording. Unless you have unusual levels of self-confidence or musical training, your reaction to this idea is to be pre-emptively embarrassed about the way your recorded self would sound. For a musician, of course, the whole idea of listening to a recording is that they are forced to confront the music they are actually making, not the idealized version playing in their head.

The idea of making and critiquing a recording translates naturally to giving presentations, and I recommended it in Chapter 4 as a tool to improve your talks. It is more difficult, although no less important, to accomplish the same goal with your writing. At the level of individual sentences and paragraphs, try reading your draft text aloud. If the wording sounds awkward or convoluted when you read it aloud, it probably is. If you need to take several breaths while reading a sentence then that sentence is very likely to confuse your readers. If you find your attention wandering by the end of a draft paragraph, then perhaps your future readers will feel the same way.

When the ideas for your work have been in your head for months, it is genuinely difficult to look at your writing as a reader will. One useful approach is to put your manuscript aside for some time, a few days or a week, for example, then sit quietly and read it from top to bottom. When doing this, do your best to think about the overall flow of your writing ("can someone understand the data in Figure 2 without the calibration information that wasn't introduced until Section 5?") and not fine-grained detail ("is that the right spelling for the Schwabmann effect?").[8] Think about giving a talk based on only the information in your manuscript in the order it is given in the manuscript. Are there ways you would want to change to order to make the talk clearer? Or is there information you would want to add or subtract? It is also useful to articulate who the readers are that will be reading your work. Scientific papers always assume some level of sophistication in their readers. The opening paragraphs and the previous literature that is cited there should clearly delineate the "background knowledge" that is assumed for a reader to find the document useful. A common mistake in writing technical papers is that the introductory text creates expectations for readers that a particular issue will be the focus of the paper only for the author to veer

---

[8] If you are reading to check fine-grained detail, starting at the end of the document and reading paragraphs or sentences in reverse order can be helpful, albeit tedious.

in a different direction later in the text. After reading only the introduction to your manuscript, could a typical reader accurately describe what the core motivation and outcomes of your work are?

A second powerful way to be honest with your writing is to get feedback from actual readers. Ask a handful of trusted colleagues who model as closely as possible the actual audience for your writing to read and critique a polished draft. If you are writing a scientific paper, then a colleague with detailed knowledge of your subdiscipline is ideal. If you are writing a research proposal, your actual reviewers may be drawn from a diversity of technical backgrounds, so finding a reader who isn't an expert in your area may be valuable.[9] The ground rules for seeking feedback like this are simple. Don't ask for enormous amounts of your colleagues' time – they should only be expected to look at one draft. Be willing to reciprocate with your own time by offering to read their work. Most importantly, listen thoughtfully and non-defensively to input. You don't have to make every change that a friend suggests, but you must at least pause and think it through. Finally, in addition to making changes to address the specific issue identified by your sample reader, think about the origin of this issue and how you can improve upon it more generally in your future writing. Thoughtful comments from a colleague who is committed to your success are a valuable commodity – don't waste them!

Be Optimistic

Learning to play music at an advanced level is hard work. There is no shortcut that can replace the thousands of hours of concentrated practice that incrementally develop a musician's skills. Heany's advice to musicians as they contemplate this situation is "Optimists always get further than pessimists or realists…Start each practice session with the belief that you are going to make progress". His advice is carefully calibrated to focus on "progress", something that can be achieved in a single practice session, not "completion" or "mastery". This is also good advice for approaching your technical writing. You can choose to begin each writing session by being confident that you will complete some action that will incrementally move your writing project forward. Alternatively, you can gaze morosely at your computer screen and tell yourself that the work will never end. The former attitude is much more helpful and has the added benefit that your friends and coworkers are more likely to enjoy spending time with you.

---

[9] For this book, I invited all of the PhD students and postdocs in my academic department to be anonymous reviewers. Each person that volunteered read a couple of chapters in draft form.

The idea of choosing to be optimistic about your progress highlights a tremendous advantage of the "schedule it and stick to it" method discussed in the first part of the chapter. As you start a writing session, remind yourself of the positive outcomes that the writers in Boice's study experienced simply by scheduling 50 writing sessions and sticking to them. Once you have your own track record of scheduled writing, you can think back to the multiple writing sessions you have completed in which you made tangible progress.

Be Persistent and Be Consistent

Heany offers wonderful advice to musicians as they think about what it takes to practice effectively and improve their skills: "Persistence is just making a plan…and sticking to it". Knowing how to be persistent doesn't mean it is easy. Heany notes that "Persistence is arming yourself with intelligence, optimism, imagination, curiosity, a sense of humor…and then applying relentless effort". Being an effective writer is hard – a willingness to put in "relentless effort" comes with the territory.

How often should a musician practice? Heany offers a simple answer: "Practice works best if you do it every day". His strong recommendation is to be consistent, where "being consistent means you have a plan, you stick to it, and you do it every day". Does this sound familiar? It should. Heany's advice encapsulates the same core idea as Silvia's description of effective writers ("Prolific writers make a schedule and stick to it. It's that simple".) and to the pattern imposed by Boice on his study subjects. If you take control of your time by scheduling time to write, your productivity as a writer and your satisfaction with the writing process are likely to soar.

Being an effective writer is not an optional skill for someone interested in a career in research; it is a non-negotiable necessity. By approaching your writing projects persistently and consistently, you can make steady progress towards your goals. The ideas we have covered in this chapter can help you on this path.

Chapter Summary

- Anything that contributes to completing a writing project counts as writing. Writing should not be an afterthought in doing technical research. Instead, think about writing early and often to help focus your work.

- Prolific writers make a schedule and stick to it. It's that simple.

- Don't wait until you "feel like it" to write. Waiting for inspiration before writing is a recipe for low productivity.

- Split a writing task into many sub-goals that can be achieved in relatively small blocks of time.

- When considering the logic of your work and its written presentation, tackle the most challenging topics first instead of leaving them for later in the hopes they will magically solve themselves.

- Try to evaluate your writing from the perspective of your intended readers. If possible, get input from knowledgeable colleagues.

- Remember that *everyone* finds writing difficult. Take pleasure in reaching small milestones instead of being discouraged by how long writing takes.

Additional Resources for Chapter 5

The key ideas in this chapter are adapted from Paul J. Silvia's book *How to Write a Lot: A Practical Guide to Productive Academic Writing* (APA Lifetools, 2007). If you only read one book about academic writing, read this one.

The biographical details describing Jim Lehrer's extraordinary "side job" as a writer come from his obituary in the *New York Times* on January 24, 2020.

Stephen King's *On Writing: A Memoir of the Craft* (Hodder and Stoughton, 2012) is a classic of the "how to" literature about writing by a master writer.

The numerical data on the value of scheduling writing came from Robert Boice's *Professors as Writers: A Self-Help Guide to Productive Writing* (New Forums Press, 1990), which also contains many other useful insights into the internal struggles most researchers have about writing.

The Elements of Style by William Strunk Jr. and E. B. White (4th Ed., Pearson, 1999) is a brief handbook on using language in writing that is known almost universally simply as "Strunk and White". E. B. White also wrote the famous children's book *Charlotte's Web*.

If you play music of any kind or are interested in how being a musical performer can suggest ideas for other creative pursuits, *First, Learn to Practice* by Tom Heany is a valuable starting point.

6

The Secrets of Memorably Bad Presentations

Chapter 5 focused on writing, arguing that writing clearly and efficiently is a cornerstone of being a successful researcher. Writing is hard, but at least you can carefully review and polish your work through multiple drafts. The same can't be said for the other vital form of communicating research, giving talks. Giving a talk has a strong element of live performance – when something goes wrong, it is there for everyone to see. No wonder that many people find public speaking nerve wracking. One widely quoted statistic says that more people are afraid of public speaking than are afraid of death, spiders or heights. Regardless of whether you enjoy, dislike or outright fear public speaking, it is hard to get ahead in most technical fields without doing it regularly.

All researchers aim to give good presentations. Giving compelling and lucid talks can help your career soar. But I am sure that you have had to suffer through conference talks or seminars that have been baffling, confused or even irritating. Giving a talk like this can have very negative effects on your career, especially at critical junctures like job interviews and talks in front of leaders in your field who will evaluate your grant proposals or tenure package.

I am going to assume that you already know how to avoid what might be termed "elementary errors" in giving technical talks. These include talking so fast that everything blends together for the audience, talking...much... too...slowly...with long...umm...err...pauses, talking too softly to be heard, TALKING TOO LOUDLY, reading text line by line exactly as it is written on a slide, always standing with your back to the audience, swinging a laser pointer across the screen and through the audience like Luke Skywalker's light saber throughout your talk, having visually distracting mannerisms and wearing wildly inappropriate or distracting clothes. In Chapter 4, we looked at one highly effective way of detecting these kinds of errors, namely, making a video of yourself giving a talk and watching it. There are other errors to avoid that are also hopefully obvious. Don't get drunk before your talk (I have seen this more than once at serious academic conferences and the outcome is never good). Don't make insensitive or insulting comments about specific people in the audience or general populations of people.

In this chapter, we will explore what might be viewed as "black belt errors" in technical presentations. These are problems that can ruin talks given by people who are highly articulate and who have great results to share. The intent of this chapter is to show you how to give presentations that aren't just slightly below average, but are so memorably bad that audience members will talk about them for years. I am not, of course, recommending that you actually attempt to give talks of this quality. My hope is that by illustrating a sizeable collection of missteps in a single place, I will help you to see ways to improve your future presentations.

A Sample Presentation

The following fictitious transcript comes from a presentation by a speaker we will refer to as David Sholl at *Engineering Applications of Molecular Crystals XXVIII*, the latest in a decades-long series of workshops that annually brings together a tight knit research community. As in many, although not all, technical fields, the speaker showed a series of slides while giving their talk extemporaneously (i.e., without notes). Some of the detailed technical content has been removed in editing for clarity; these gaps are marked as [...]. As a contributed talk, this presentation was allotted 15 minutes, including time for questions, on the afternoon of the third full day of the workshop. The preceding days of the workshop featured plenary and invited talks from several luminaries in the field and 15–20 contributed talks each day. Because of the setting and format, the audience for the talk were experts in the technical topics being discussed and were hoping for a concise description of an original piece of technical research.

This fictional talk imagines an event with a live audience, but the suggestions drawn from it apply equally well to giving talks virtually. A series of highlights have been included to point out choices that made the presentation bad and hopefully help you to avoid similar missteps in your future presentations. Without further ado, let's hear from our speaker....

Good afternoon everyone. Is this microphone on? Oh, I (*inaudible noise*) oops. Oh I see, that *was* on. Let's see, you just heard the title of my talk from the session chair, so I won't read it out again. As you can see from my first slide, oh, wait that's not right. Hmm, just a second ... ha ha, my mistake, that is an old version of my slides. Let me just switch over to the correct set. (*Long pause for the speaker to fiddle with their computer, opening several files and accidentally showing a page of an emoji filled chat before finding the correct presentation.*) (See Mistake 1 boxed text.)

MISTAKE # 1: WASTING THE AUDIENCE'S TIME WITH EASILY AVOIDABLE TECHNICAL DIFFICULTIES

Unless it is truly impossible, test out everything you will need for your talk (computer connection, microphone, pointer, etc.) in the specific venue for the talk well before the event actually starts. If your presentation involves showing video or playing audio, test this in the specific venue where you will be speaking. If you are giving a talk at a conference, attend one or more talks in the same room and think from the audience's perspective how well they can hear and see and adjust your presentation accordingly. As the old saying goes, you never get a second chance to make a first impression. Don't fill your first impression with technical difficulties.

As I was thinking about today's talk, I was reminded of a story about Albert Einstein when he lived in Princeton. One day Einstein got on a train in Princeton. All was well until the conductor came through the carriage asking for tickets. Einstein patted his jacket pocket, then slowly went through all of this pockets and his briefcase, but wasn't able to find his ticket. "Don't worry", the conductor said "I am sure everything is fine". When he came back through the carriage a few minutes later the conductor was surprised to see Einstein on his hands and knees, apparently looking for his ticket under the seat. Embarrassed, the conductor rushed in and said "Please stop. I know who you are. I am honored to have you on board and I am sure you have a valid ticket". Einstein fixed his gaze on the conductor and said "Young man, I also know who I am. But I do not know where I am going!". Ha ha – let's hope that isn't the situation we find ourselves in on the conference excursion tomorrow (See Mistake 2 boxed text).

MISTAKE # 2: BEGINNING WITH AN IRRELEVANT AND TIME WASTING ANECDOTE

This joke took up roughly 5% of the total time available to the speaker. There are some settings where using some aptly chosen jokes can be appropriate – a speech at a wedding or a retirement party, for example. When giving a technical talk, however, using even a few percent of your time for this purpose will give the impression that you don't have much in the way of useful results to talk about. When speaking to a multicultural audience, as is common in scientific settings, humor has the added challenge that quite a few people in the audience are likely to not get the joke.

To give some context for the results I will present today, it is useful to begin with some historical context. As we all know, people's lives have been affected by molecular crystals for thousands of years. Some of the earliest known surgical procedures were for removing kidney stones, which are mainly formed from crystallization of uric acid. Historical figures including Oliver Cromwell, Isaac Newton and Benjamin Franklin were afflicted by kidney stones. Crystallization was critical in the early understanding of molecules. Louis Pasteur's experiments with crystals of tartaric acid, a byproduct from wine making, gave the first direct evidence of chirality in chemicals, although the molecular origin of this effect wasn't really understood until many years later. Today, the active ingredients of many pharmaceuticals are molecular crystals, and controlling the crystal structure and, therefore, solubility of these crystals is a key part of the regulatory and manufacturing environment for these drugs. The specific molecules I will talk about today aren't chiral, and they certainly aren't under consideration as pharmaceuticals, but still, it is cool to see how thinking about molecular crystals connects back to the field's historical roots (See Mistake 3 boxed text).

MISTAKE # 3: USING BANAL AND PATRONIZING BACKGROUND INFORMATION

If you are giving a talk to a broad audience, by all means include some background on why your particular field is interesting. But if you are giving a talk at a conference titled *Molecular Crystals XXVIII*, it is safe to assume that your audience already knows a thing or two about molecular crystals. Like mistake # 2, spending your time going over information that your audience will think is generic will give the impression that you have little to say. If you find yourself saying, like our speaker did, "As we all know..." watch out.

The next slide shows an outline for my talk today. As you can see ... oh, wait a minute ... oops, I forgot that I animated this slide and have to click for each point. As you see, I'll start with an Introduction, then ... aah, oh there it goes ... discuss my Methods. This will lead us to the Results of the work, and then I will sum up at the end with Conclusions. That will leave some time for my ... oops, this button is tricky to see ...there we go, time for my Acknowledgements and then I can try and answer any questions you might have (See Mistake 4 boxed text). (*At this point, the session moderator tried to catch the speaker's eye and mimed looking at their watch. By this point, the speaker had used up several minutes of his allotted 15 minutes without saying a single word about the real substance of his talk.*)

MISTAKE # 4: USING AN INFORMATION-FREE OUTLINE

Unless it is very unusual setting, surely your audience will expect you to begin with an introduction and end with a conclusion without you explicitly telling them. Some templates for giving a talk advise including a slide with an outline, but I have never understood why this is necessary. Think about your favorite popular musical artist performing. Do they begin by explaining "First we will sing a verse, then the chorus, then a second verse, then the chorus a couple of times, followed by a guitar solo..."? For your next presentation, channel your inner rock star and skip the outline.

Today I am going to talk about connections between molecular crystals and zeolites, and especially what can be learned from detailed modeling of zeolite crystals. Zeolites are used on massive scales in several industries, including petroleum refining and detergents, so they have been widely studied. As you can see from my slide (Figure 6.1), there has been a lot of work on modeling these materials and there are now many examples where the accuracy of these models is in many senses comparable to experimental data (See Mistake 5 boxed text).

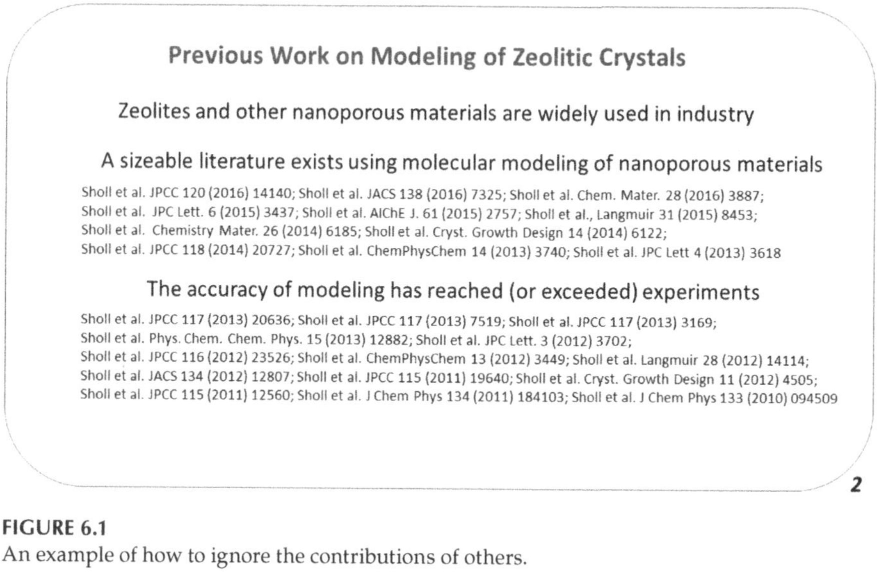

Previous Work on Modeling of Zeolitic Crystals

Zeolites and other nanoporous materials are widely used in industry

A sizeable literature exists using molecular modeling of nanoporous materials

Sholl et al. JPCC 120 (2016) 14140; Sholl et al. JACS 138 (2016) 7325; Sholl et al. Chem. Mater. 28 (2016) 3887; Sholl et al. JPC Lett. 6 (2015) 3437; Sholl et al. AIChE J. 61 (2015) 2757; Sholl et al., Langmuir 31 (2015) 8453; Sholl et al. Chemistry Mater. 26 (2014) 6185; Sholl et al. Cryst. Growth Design 14 (2014) 6122; Sholl et al. JPCC 118 (2014) 20727; Sholl et al. ChemPhysChem 14 (2013) 3740; Sholl et al. JPC Lett 4 (2013) 3618

The accuracy of modeling has reached (or exceeded) experiments

Sholl et al. JPCC 117 (2013) 20636; Sholl et al. JPCC 117 (2013) 7519; Sholl et al. JPCC 117 (2013) 3169; Sholl et al. Phys. Chem. Chem. Phys. 15 (2013) 12882; Sholl et al. JPC Lett. 3 (2012) 3702; Sholl et al. JPCC 116 (2012) 23526; Sholl et al. ChemPhysChem 13 (2012) 3449; Sholl et al. Langmuir 28 (2012) 14114; Sholl et al. JACS 134 (2012) 12807; Sholl et al. JPCC 115 (2011) 19640; Sholl et al. Cryst. Growth Design 11 (2012) 4505; Sholl et al. JPCC 115 (2011) 12560; Sholl et al. J Chem Phys 134 (2011) 184103; Sholl et al. J Chem Phys 133 (2010) 094509

2

FIGURE 6.1
An example of how to ignore the contributions of others.

MISTAKE # 5: IGNORING THE
CONTRIBUTIONS OF OTHERS

In the slide in Figure 6.1, the speaker managed to ignore all work by people outside of his group. He also didn't mention by name any of the many other people who were involved in the papers that did get listed, a choice that demonstrates a high level of skill with this particular mistake. Being generous with credit to other people is a guaranteed way to make them feel good. Making connections with other work in your field also demonstrates the breadth of your knowledge. You should aim to show that you are generous towards others and genuinely open to outside ideas.

To give a sense of the kinds of molecular models we have been using in our work, my next slide shows a couple of examples (Figure 6.2; See Mistake 6 boxed text). The technical details and convergence aspects of these calculations are quite important, as I have indicated on the slide (See Mistake 7 boxed text). Having shown these initial examples, I will move on to my main topic for today.

MISTAKE # 6: USING MANY FONTS
ON THE SAME SLIDE

The slide in Figure 6.2 packs eight different fonts into a single distracting page. This is likely to leave viewers with an uncomfortable feeling that there is some underlying message or logic to these choices that they can't understand, or at least the sense that the slide was hurriedly cut and pasted from many different sources. Similar effects can be achieved with excessive and arbitrary use of colored text, *not to mention italics* <u>and underlining</u>.

MISTAKE # 7: USING AN ABUNDANCE OF
UNEXPLAINED JARGON AND ACRONYMS

The speaker's slide, which was shown to the audience for only about 20 seconds, is jammed with abbreviations and technical terms. This kind of language can be appropriate in a written report that is designed to give precise details, but unless text can be read and understood by the audience in the time they have to view a slide, the value of the text in a presentation is questionable.

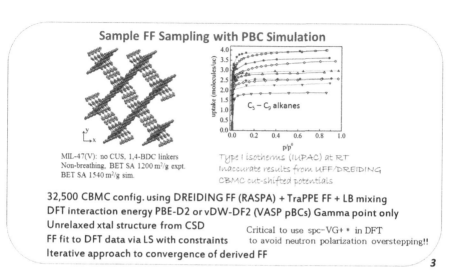

FIGURE 6.2
How not to choose fonts. This slide was only shown to the audience for about 20 seconds.

Our first set of calculations focused on structural aspects of pure silica zeolites (Figure 6.3). There is a lot packed in here – I hope you can see it OK from the back, ha ha – but the message is clear (See Mistake 8 boxed text). You will especially notice the two examples I highlighted at the bottom of the slide, for structure codes ITW – oops, I mean IHW, that font is really quite small even from up here near the screen – and LTA. These two structures stand out from the rest (See Mistake 9 boxed text), but that is probably because of the non-orthogonal low symmetry distortions than can occur in them at moderate conditions [...]

MISTAKE # 8: USING FONTS AND SIZES THAT CAN'T BE SEEN EASILY

You will know you have made this mistake if you ever feel the urge to apologize, as the speaker did, for things that are hard to see in a presentation. Similar to Mistake # 1, a judgment about what is easy to see needs to be made from the point of view of an audience member sitting in the actual venue for your talk.

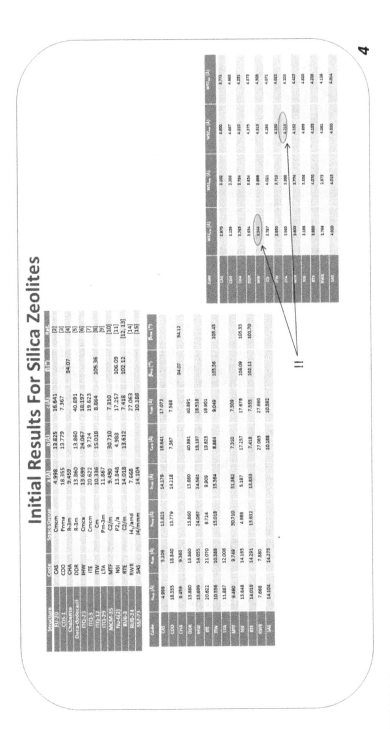

FIGURE 6.3
An example of how not to use tabulated data on a slide.

MISTAKE # 9: SHOWING A SLIDE WITHOUT ANY TEXT OR CONCLUSION

Glance quickly at Figure 6.3, then try and describe out loud what the main point of the slide is. Because the speaker didn't use any kind of text on the slide, it is impossible to know what he wanted listeners to remember. A good rule of thumb is that almost every slide should have a brief summary statement giving your audience a "take away message". If you have trouble crafting a statement like this for a particular slide, this is a strong hint that you haven't planned this section of your talk clearly.

After our initial tests with pure silica materials, we extended our efforts to a diverse set of more complex compositions. On this slide (Figure 6.4), I want you to focus on the results in part (g) of the figure, particularly the small "knee" you can see in the uptake data (See Mistake 10 boxed text). If we saw this kind of thing in experimental data, it might hint that there are kinetic effects in the experiment, but remember that these results are coming from simulations that avoid these possible non-equilibrium effects. This means that the knee is in fact a signature of some underlying heterogeneity in the specific structure used in this example. Although I didn't include the data on this slide, we have also seen evidence for this kind of thing in several other cases […]

MISTAKE # 10: INCLUDING (LOTS OF) DATA THAT ISN'T DISCUSSED AT ALL

In addition to showing the audience another slide with unreadably small text and figures, the speaker shows a figure with 12 sub-figures and then completely ignored 11 of them. Complex multi-part figures like the one shown in Figure 6.4 are useful in some publications that have strict limits on the total number of figures, but that is no excuse for using them in a presentation. Figure 6.4 also repeats mistake # 9.

(We rejoin the talk just as the speaker has talked for about 10 minutes of their 15 minute total.) The work that I have presented so far gives as a strong hint that something very interesting is going on with the new materials I mentioned. Hmm, I can see that my time is a little short, so I think I will skip this slide, and this one. And I can also jump over these three slides. And this one

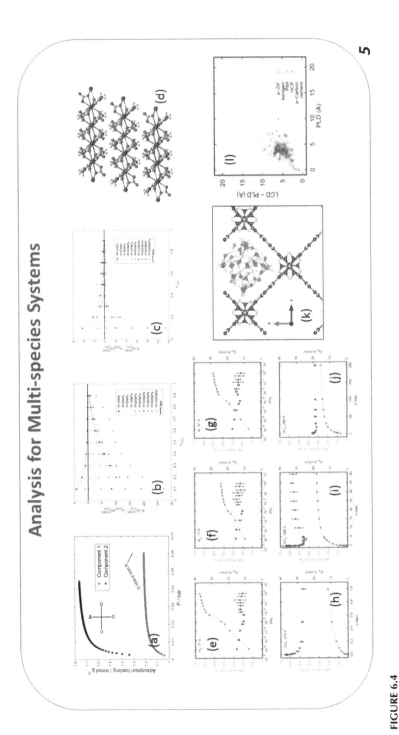

FIGURE 6.4
If a picture is worth a 1000 words are 12 pictures worth 12,000 words? An example of unnecessary data overload.

too (See Mistake 11 boxed text). Actually, let me back up a couple of slides. I won't talk about this in detail (See Mistake 12 boxed text), but will just point out that the simulated vibrational spectra you can see here show several unconventional band crossing points [...] (*This technical explanation continued for another 90 seconds.*)

MISTAKE # 11: SKIPPING SLIDES BECAUSE OF LACKING OF TIME

It is often a good idea to include a few backup slides that address particular questions you anticipate about your work. Having sections of a presentation that get jettisoned "on the fly", however, communicates to the audience that you couldn't be bothered to carefully plan how you would use their valuable time.

MISTAKE # 12: SKIPPING SLIDES BUT THEN TALKING ABOUT THEM ANYWAY

Skipping slides at least gives the audience a sense that you are somewhat aware of their time constraints. If you really want to get under the audience's skin, let them know that you are running short on time, say you are going to skip over some topic, and then give a prolonged discussion of that topic anyway.

I can tell from the background noise outside the room that people from the other conference sessions have started the coffee break. Let's hope there will be some coffee left when we get there – it was a shame they ran out yesterday, ha ha. Let me give you a summary of my talk (Figure 6.5). We have been very pleased with how well our modeling methods can describe a whole range of experimental data. This has convinced us that we can use these models to discover new materials in advance of experimental characterization and we have already seen some early successes with this idea. I had to skip over a couple of the examples of this, but you can of course ask more about them in the questions.

That brings me to the next topic I'd like to touch on today (See Mistake 13 boxed text). As you can see on this slide (Figure 6.6), we have started applying the same ideas to partially disordered materials (See Mistake 14 boxed text). There are some challenging technical aspects to doing this, so I'd like to take just a couple of minutes to talk about them [...]

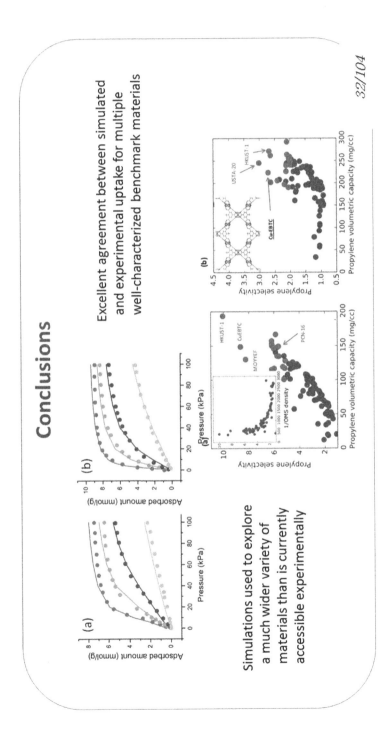

FIGURE 6.5
Finally, the end of the talk is in sight. Or is it?

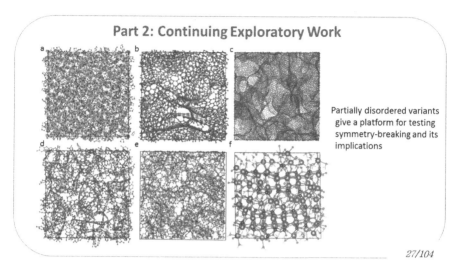

FIGURE 6.6
A slide that lets the audience know they have no idea how long the talk is going to go for.

MISTAKE # 13: CONFOUNDING AUDIENCE EXPECTATIONS WITH A FALSE ENDING

Our speaker gave the audience multiple cues that he was about to stop talking (a slide titled "Conclusions", a comment about the upcoming food break, a "summary" of results). No matter how fascinating your talk is, most of the people in any audience you speak to are looking forward to your talk ending. If you are not about to finish your talk, be extremely cautious about giving any cue that suggests you are. I have been asked by many speakers visiting my academic department how long their seminar talk should be. My stock reply is "When was the last time you went to a talk that finished early and you were disappointed?".

MISTAKE # 14: USING SLIDE NUMBERS TO INJECT UNCERTAINTY INTO THE AUDIENCES' MINDS

Using slide numbers is typically a good idea. It is much easier for someone to start a question with "On slide 12..." than "Can you find that slide with the pink and yellow structure?". But did you notice that the speaker's slide numbering scheme changed between Figures 6.4 and 6.5 and that the numbers on Figure 6.5 suggest there are 70+ more slides still to come? Even worse, the slide numbers went backwards as the speaker advanced from Figures 6.5 to 6.6.

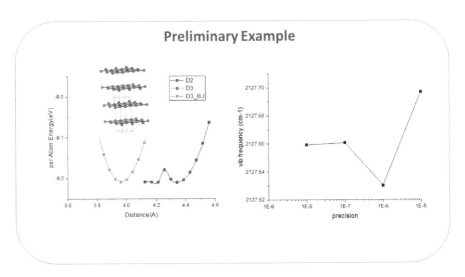

FIGURE 6.7
"We just generated this data overnight…".

As an example of this approach, here is some data that is "hot off the press" as they say (Figure 6.7). My student just emailed this to me this morning from some calculations he did overnight, so we haven't had a chance to look at it too carefully. In fact, the data on the right looks a bit odd – hmm, perhaps some kind of calibration issue (See Mistake 15 boxed text). In any event, what we think this will let us do is […] (*At this point, the session chair stood and pointed emphatically at their watcher as the speaker continued.*)

MISTAKE # 15: SHARING RESULTS YOU HAVEN'T THOUGHT ABOUT CAREFULLY

In some very informal settings, such as within a research group, presenting brand-new data before you have thought about it can be appropriate and useful. In a formal presentation where you are presenting accumulated work from months or years of work, throwing in "last minute" data will deeply undermine your audience's confidence.

Let me acknowledge the people who contributed to this work: first, of course, the PhD students and postdocs who have worked in our group over many years. I have highlighted on the slide the names of people whose results I showed today, and I especially want to point out Dr. Lincoln, who implemented several of the key methods I showed you in the main part of the talk. We also have benefitted greatly from assistance over the years from the technical staff at the Southeastern Computational Center and Data Repository, and they like me to mention the financial support they get from the National Science Foundation. Some of the initial ideas for this work were motivated by discussions in the Poland-Lithuania-Georgia membrane working group, and I appreciate funding from the Department of Energy that allowed me to be an observer at those meetings over the years. I didn't show the data today, but we have also had help in some of our structured search approaches from students in a course I teach, ChBE 4207, so thanks go to all the students who have been in that class. We received financial support from the Department of Energy Office of Science and also the Advanced Manufacturing Office under the grant numbers listed on the slide. Recently, we also received funding from the Department of Homeland Security as part of their GNOME-HAT program, as well as the NIH via the Rat Survival Initiative Director's Exploratory Scheme. And I can't forget to mention several industry partners who are working with us on applied aspects of this project, the LiquiPure corporation, the R&D arm of GTR Inc., and also a small sponsorship from Uncle Bob's Septic, who help keep things in shape at my summer cabin and make our annual group retreat there much more pleasant (See Mistake 16 boxed text). And now I think it is time for some questions.

MISTAKE # 16: DRAINING THE AUDIENCE'S ENTHUSIASM WITH EXTENSIVE ACKNOWLEDGEMENTS

A key part of any scientific paper is a section that carefully acknowledges financial support and intellectual contributions to the work. These acknowledgements should also be given in talks, but there is no compelling reason to put them at the end of the talk. Just as first impressions are important, the end of events has a disproportionate impact on people's impression of the event. For this reason, I *strongly* recommend that you cover the acknowledgements for your talks at the beginning, right after the title slide, and think carefully about ending with a strong slide that shows the audience the key items you want to remember. This may sound like a simple change to make, but I suspect if you try it once or twice, you will never go back to the typical "end with acknowledgements" order.

(*A question from the audience*) Can you clarify where you get the structures for the materials you model? It seems like that is a necessary input for each calculation.

Yes, we have to be careful with this. Your question reminds me of something quite different but also important (See Mistake 17 boxed text). I showed some data from biphasic systems at the beginning of the talk, but we found that the convergence of our calculations was slow for these systems unless we used overdamped basis functions for [...]

MISTAKE # 17: ANSWERING A QUESTION WITH A TANGENT TO ANOTHER TOPIC

If an audience member asks you a question, it is usually because they would like an answer. Answering a completely different question may be a common strategy for politicians, but it won't leave your audience feeling that you respect their intelligence.

(*Another audience question*) I have a simple yes or no question. Did you include Brockman relaxation in your refinement of disordered structures?

As you know, I'm sure, disordered structures create all kinds of challenges, and we are still developing some tools to better tackle them (See Mistake 18 boxed text). One way to start with this is to take a boundary-free approach, but we have found that this brings in a whole new set of complications so we tend to avoid it. Instead, we have lately [...]

MISTAKE # 18: ANSWERING A SIMPLE QUESTION WITH A LONG DISCOURSE

If a question can be answered with a single word or a single sentence, give that answer and stop! By the question and answer period of almost any talk, most of the audience is itching to get on to other things. Don't use every question as a springboard for another monologue.

(*A final audience question*) Real-world applications of these materials often ...
Yes, I see where you are going with this (See Mistake 19 boxed text). We didn't really consider cost-at-scale questions so far, but as we move forward [...]

MISTAKE # 19: CUTTING OFF A QUESTION BEFORE IT IS ASKED

Listening attentively is a basic way to show respect to others. Don't cut people off, shake your head dismissively or otherwise communicate impatience when you are asked a question after a presentation.

(*Final comments from the session chair*) Well we are quite a long way over our scheduled time now, so let's hope there are still some refreshments left. Please join me in thanking our speaker.

Chapter Summary

There are many pitfalls in giving technical presentations. I suspect that you have seen many of the mistakes listed above in talks and seminars you have attended. Perhaps you have even committed a few of them yourself. To help you remember the many ways that things can go wrong, use the game of *Bad Talk Bingo*. Use the "bingo card" in Figure 6.8 as you sit in future talks and cross off categories as the speaker fulfils them. If you can complete a set of five squares horizontally, vertically or diagonally you "win", but please don't jump up and yell "Bingo!".

Giving great talks is hard. As you gain experience and practice your self-confidence and ability in giving talks will improve. As you do, I hope that by learning from the mistakes of others you can avoid the experience of giving memorably bad presentations.

Reading line by line	Excessive "umms" and "arrrs"	Lengthy credits & thanks	Skipping slides (then talking about them anyway)	Question cut off before answering
Information-free outline	Talking too fast	No words on slide with much data	Introduction too general/banal	Talking too quietly
Tangential answer to question	Technical difficulties	False ending	Talking too slowly	Long answer to simple question
Talking too loud	Back always to audience	Irrelevant anecdote	Ignore contributions by others	Talk too long
Presenting recent unanalyzed data	No clear conclusion	Skipped slides	Banal/obvious conclusions	Excessive acronyms and jargon

FIGURE 6.8
The Bad Talk Bingo card.

Further Resources for Chapter 6

All images and data on the slides from the fictitious talk described in this chapter come from internal work within my own research group, but they have deliberately been taken out of context and presented poorly. Some of the technical content on the slides and in the talk is based in fact, but some was simply made up following the best *Star Trek* traditions of using technically sounding jargon to make things sound impressive. No blame should be assigned to the graduate students and postdoctoral researchers who actually did the underlying technical work; they would never dream of showing their data in the confusing ways I chose to use above.

A live version of a talk that illustrates many of the same problems discussed in this chapter is available by searching YouTube for a video with the same name as the chapter.

The concept that endings have a disproportionate influence on audience perceptions (see Mistake #16) came from the studies of this phenomena described in Daniel Kahneman's *Thinking Fast and Slow* (Farrar, Straus and Giroux, 2013).

7

How to Win the Nobel Prize and Change the World

This book has aimed to point you towards skills and habits that can make your creative research career more successful and, just as importantly, more satisfying. The motivations people have for pursuing research vary widely. It is rare, however, for researchers to devote months or years to a problem that they know is obscure and to hope that no one in their professional community or elsewhere will ever care about their results. Instead, almost every researcher hopes that someone somewhere will care about their work and perhaps even use their results in a positive way.

The ideas in this concluding chapter are adapted from the book *A Beginner's Guide to Winning the Nobel Prize* by Peter Doherty, an Australian immunologist who shared the Nobel Prize in Medicine in 1996 for his work on how our immune systems recognize and act against viruses. The chapter title is a deliberately inflated version of Doherty's tongue-in-cheek book title. In his book, Doherty points out that he can't actually tell anyone how to win a Nobel Prize, although (unlike me) he can at least talk about the topic from firsthand experience. Hopefully, you will keep reading even with the caveat that the title of this chapter shouldn't be taken too literally.

Peter Doherty was born in 1940 in Brisbane, Australia. After attending his local public high school, which shared a tongue-twisting name with the suburb of Indooroopilly where it is located, Doherty studied at the local university to be a vet. Developing an interest in research, he moved to Edinburgh in the UK to complete a PhD. In 1970, he returned to Australia to do postdoctoral research. It was this work, performed within just a couple of years of his PhD, that won Doherty and his postdoctoral supervisor, Rolf Zinkernagel, the 1996 Nobel Prize. Just like Fiona Woods, who we met in Chapter 1, Doherty was chosen as Australian of the Year, in his case in 1997.

Doherty's book and the title of this chapter use the idea of a Nobel Prize as a shorthand for work that has long-lasting significance and meaning. In scientific publishing, the term "impact" has taken on a narrow meaning associated with the number of times a paper is cited in a short period after it is published. Many scientists use the resulting "impact factors" to help decide which journals they want to publish in. Instead of fixating on the number of citations a paper gets, I urge you to think about the impact of your work using the everyday meaning of this phrase. Will your work change how people in your field think about something? Will your work be useful in some way to people outside the tiny group of people studying your specific problem? Does your work spark creative connections that allow someone to solve a related but different problem? These kinds of questions are the real measure of impact, and they point to impact being hard to quantify and taking place over long periods of time. The more than 20-year delay between Doherty's postdoctoral work and his Nobel Prize is one example of this time lag.

The discussion above about significance and impact centered on achieving these goals in your research. There are also many professional and personal activities that are significant and impactful without being centered on technical outcomes of research. Your research might train undergraduate students in careful work and logical thinking, skills they go on to use in a broad range of careers. You might devote your energy and talent to advancing education or life prospects for a community who is traditionally disadvantaged. Your training might allow you to become involved in influencing local or national government policy. All of these areas, and others you can probably add to the list, have the possibility of leading to real significance and impact. Many of the research-motivated ideas we explore in the rest of the chapter are also highly relevant to non-research endeavors.

Try to Solve Major Problems and Make Big Discoveries

Doherty's first piece of advice seems almost tautological: if you want to make a lasting impact with your research, work on an important problem. This advice is not just a Zen Koan from a master researcher; it can be a helpful way to frame the choices you must make about problems you could work on. This advice forces you to think about what makes a problem important. One way to address this issue is to imagine that your specific research project is completely and wildly successful. What would this look like? Whose lives or future work would be influenced by your research results? If this daydreaming exercise convinces you that your project has enough potential upside to be worthwhile, then it is also healthy to imagine what problems could stand in the way of that success. If the most positive outcomes you can dream up even as a wild-eyed optimist seem limited, perhaps the project isn't attacking what Doherty calls a "major problem".

In the early 1980s, it was accepted wisdom that ulcers were associated with high-stress hard-charging lifestyles. To give just one pop culture example, one of the Madison Avenue executives in *Mad Men* is depicted as suffering from an ulcer, which he and his doctors assume is from his lifestyle. In 1982, two doctors at a hospital in Perth, Western Australia, Barry Marshall and Robin Warren, had a startling idea; they suggested that ulcers were caused by a bacterial infection. Let's evaluate the value of doing research on this idea from the "importance" metric outlined above. If the research was wildly successful, there is no doubt that many people would benefit, so from this perspective, it was potentially a worthwhile problem. At the same time, however, finding something that overturns many decades of medical wisdom seems exceedingly unlikely. Marshall and Warren submitted a paper with some preliminary results to an obscure Australian medical conference. The paper's reviewers ranked it in the bottom 10% of papers for the year, and it was rejected.

Despite the initial lack of enthusiasm for his work, Marshall remained convinced that it was worthwhile exploring the connection between bacteria and ulcers. In 1984, he performed an experiment that surely would not be approved by any present-day ethical review board by drinking from a Petri dish in which he had cultured *Helicobacter pylori* (*H. pylori*). Within three days, he developed nausea and his mother noticed another symptom of digestive problems, halitosis. Just a few days later, Marshall experienced vomiting and severe stomach inflammation. After waiting another week and enduring an endoscopy that showed *H. pylori* had infected his stomach, Marshall started taking antibiotics. This dramatic experiment gave direct evidence that gastritis could be caused by bacterial infection, that is, that gastritis was an infectious disease. This insight revolutionized the treatment of gastritis and ulcers. In recognition of their work, Marshall and Warren were awarded the Nobel Prize in Medicine in 2005.

Medical breakthroughs such as the work of Barry Marshall and Robin Warren are appealing in part because their impact is easy to understand. Most research, including most biomedical research, has impact is less obvious ways even when it is successful. Doherty's advice to "solve major problems" can nevertheless be helpful in setting long-term goals for your work. Like "write my thesis", "solve a major problem" is not a useful item to put on your daily to-do list. As with any significant long-term goal, breaking the overall task into well-defined actions is critical to connecting your ambition to make big discoveries with the day-to-day realities of creative research.

Acquire the Basic Skills and Work with the Right People

In Chapter 5, we explored some analogies between learning a musical instrument and effective technical writing. No one can become a concert pianist without enormous amounts of diligent practice, lessons with gifted teachers and experience playing music with high-quality ensembles. Technical skill in playing a musical instrument isn't the same as making beautiful music, but without the necessary skills, musicians simply can't participate in the top level of their profession. The same observations apply to being a highly accomplished researcher. In addition to generic skills such as thoughtfully reading the literature and critical thinking, every research field requires a deep set of technical skills. Researchers who deeply understand the strengths and limitations of their experimental equipment will go much further than those who view equipment as a "black box" that delivers data that can't be questioned. You should become a connoisseur of the detailed techniques that underlie your research and continually add to your basic skills.

In most fields of technical research, a key period of acquiring skills occurs during completing your PhD and, in many cases, postdoctoral research. Doherty points out that to have long-term success in research, performing this training with the "right people" is vital. Like his advice to work on "major problems", this suggestion is helpful because it forces us to think about what characteristics the "right" research mentors ideally have. Because of the community aspects of research being successful, it is useful to have a mentor who is well-connected in the sense that they know the state of the art and the history of their field. A mentor with these qualities is not necessarily someone who follows the latest fads in their own work, but they will at least understand what will be necessary if you undertake a project that swims against the prevailing currents.

You can (and should) learn about the technical work of a prospective mentor by reading their papers. If possible, you should also talk with current and previous students from the department and research group you are considering joining. Ask them about the intellectual and personal environment, how credit is shared, and how people go about learning new techniques. Working with a powerful, well-funded research mentor might not be a positive experience if you rarely see them and fellow group members jealously guard their work to avoid being "scooped" by someone else. Alternatively, an award-winning world-famous researcher might be an inspiration to everyone in their research group and work closely with each person to further their success. Understanding where on the spectrum between these two extremes a potential mentor lies is not possible just from reading their work.

Choosing a research mentor based primarily on the ranking of their institution or their "fame" in the research community is almost always a bad idea. The various rankings that exist of research institutions have strong limitations that often favor the "name brand" status of top institutions much more heavily than the specific skills of individuals at those institutions. A correlation usually exists between ranking and the overall resources available at an institution, but only in a coarse-grained way. The quality of the specific individuals you work with is likely to be much more influential than the name of the institution where you work. Peter Doherty himself is a good example of this situation, getting his undergraduate degree at the University of Queensland, his PhD at the University of Edinburgh and then doing his Nobel Prize winning postdoctoral research at the Australian National University. Each of these places are respectable institutions, but at the time Doherty was educated in the 1960s, none of them would have appeared high on any international ranking of elite research universities.

As you explore options for your own training and early career, don't become paralyzed by concerns that you have to find a "perfect" location. Many people face constraints on where they can work for family, visa or other reasons. To return to our analogy about musical performers, skilled teachers and colleagues cannot overcome the disadvantages that will be faced by a musician who doesn't practice diligently. Among the factors that will help you acquire the basic skills to move your research forward, your own attitude and your willingness to work and think are the most important. There are almost certainly many different locations and mentors around the world that can contribute to your success.

Talk about the Problem

Let's say you have followed Doherty's advice so far to the letter. You have chosen a Big Important Problem to work on. You have trained with skillful, dedicated mentors and have become technically virtuosic. Almost certainly you will soon realize that one reason that problem is big and important is that it is really hard. A project like this will often reach a stage where hard work alone won't lead to progress. Instead, some kind of new idea is needed. I would argue that this is where you need a full arsenal of good research habits if you are going to make progress. The idea creation methods from Chapter 4 are a great place to start because they can give some structure to the challenge of coming up with something new.

Doherty describes a critical strategy for keeping creativity flowing when a project risks getting stuck: talk about the problem. We saw this in a different guise in Chapter 3 as Bad Habit #3, where an ineffective researcher works alone. Take every opportunity you can find, formally and informally, to talk about your work. Having to put your thoughts into words can help you to identify places where you are thinking too glibly or have doubts. This is especially true if you are able to find colleagues who will listen and question you when you attempt to explain what is going wrong with your work instead of only what is going right. The way that Andrew Wiles involved a trusted colleague and later a former student in his quest to prove Fermat's Last Theorem (see Chapter 3) are examples of this approach.

Generating new ideas by talking with others is a concrete example of the value of diversity in successful research. If you only talk with people who are just like you in terms of their intellectual training and personal background, you are missing chances to tap into truly creative ideas. Making the effort to genuinely communicate your ideas to people from different fields and backgrounds and, just as importantly, to understand their questions and ideas takes time, but the potential output in terms of creativity is high.

A final note on effectively talking about your research problem with others: don't be someone that just talks about themselves all the time. Being self-centered is an unattractive trait in friends, and the same is true in researchers. If you ask your colleagues what in their work excites them and what some topical challenges are in their subfield, then actually listen to what they will have to say you will be amazed at how their opinion of your professional skills increases.

Learn to Write Clearly and Concisely

The topic of Chapter 5 was habits that can help you write efficiently. The importance of technical writing in achieving long-lasting research success is also emphasized by Peter Doherty in his list of advice. Doherty goes beyond just the idea that writing needs to get done and points to two key stylistic aspects of impactful technical writing: clarity and conciseness. In a perfect world, readers of your technical papers would evaluate your ideas without concern for how long the papers are. In reality, your readers are just as time-constrained as you are, so rambling descriptions of your work that require intense attention to understand are likely to mean that people simply don't read it. Doherty put this in an admirably clear and concise way: "Science is about telling good, readable, memorable stories".

An entertaining example of the power of brevity comes from a 2011 physics paper about the properties of neutrinos. Earlier that year a group of experimental physicists had reported data suggesting that neutrinos could travel faster than the speed of light. If this conclusion was correct, it would contradict a key assumption in modern physics, so these results generated intense interest in the physics community. By 2012, the original experimental results were shown to be in error,[1] but not before a large number of papers were written on possible mechanisms for faster-than-light travel. When multiple papers are coming out in a rapidly moving field every week or possibly each day, getting people to pay attention and remember your results can be difficult. A group of physicists led by Michael Berry at the University of Bristol faced this challenge with a paper they titled *Can apparent superluminal neutrino speeds be explained as a quantum weak measurement?* This is a well-constructed title since it immediately told readers what the aim of the paper was. The abstract of the paper is a masterclass in concision: it reads, in full, "Probably not". Is this abstract a "good, readable, memorable story"? Without a doubt, the title and abstract meet Doherty's conditions. If I want to understand the properties of faster-than-light neutrinos, then this pithy abstract tells me I don't need to worry about quantum weak measurements (whatever those are).

---

[1] Once they were found, the sources of the problems with the original experiments, which included a loose connection between two electronic components, were mundane.

I am not recommending that you trim the abstract of your next paper to two words. It is worth remembering, however, that many potential readers of your work might only look at the title or just the title and abstract. Does your title clearly convey the core aim of your work? Does your abstract tell readers the key conclusion from your work that you want them to remember? If not, or if you have a hard time articulating what these aims and conclusions should be, it would be wise to write them again from scratch after pondering Doherty's dictum of presenting your work in a form that is readable and memorable.

It is easy for me to say you should write clearly and concisely, but how can you learn to do so? As with other skills underlying excellent research, being able to write well is a learned skill that develops through attention and practice. Don't feel discouraged if you aren't currently confident with your writing. Instead, do your best to focus on ways you can make improvements. A powerful way to do this is to adapt the methods Benjamin Franklin used as a young man in the early 1700s to become a better writer, long before he went on to become an accomplished scientist and one of America's founding fathers. Franklin, who had limited formal education, took articles from the literary magazine *The Spectator*, read them and took notes. A few days later, Franklin used his notes to write his own version of the articles. He then carefully compared his writing to the original with the aim of finding weaknesses in his own writing. You can readily adapt this approach to improve your own technical writing: find examples of writing that you or someone whose judgment you trust consider "good", summarize a few key paragraphs of the writing in bullet point form, then wait a few days and write your own version of the paragraphs from the notes. Finally, critically compare your writing to the original text. Doing this is hard work, to be sure, but the long-term payoff can be significant.

The "Franklin method" addresses the important but fuzzily defined goal of writing well. You can also practice the more specific skill of writing concisely by editing what others have written. Take a few sentences or a paragraph from a technical source and write a new version with the sole aim of reducing the length without changing the meaning.[2] Can you reduce the number of words by 10%, 20% or more? Once you develop a knack for this kind of editing, apply it to your own writing to "clean up" your work once you have a nearly complete draft of your content.

Focus, Don't Be a Dilettante

In 2020, Michael Hawley died of colon cancer after living an extraordinary life. Hawley worked for Lucasfilm and then Steve Jobs in early 1980s, wrote famous commencement speeches for Steve Jobs and also Larry Page of

---

[2] In addition to improving your general writing skills, this will help you when writing research proposals and similar documents that have strict length limits.

Google, was the scientific director for an expedition to Mount Everest and directed research projects that anticipated what is now called the Internet of Things. He also won first prize in the most prestigious international competition for amateur pianists, the Van Cliburn Competition, playing his own arrangement of music by Leonard Bernstein in the final. He later played with famed cellist Yo-Yo Ma at the wedding of Bill "The Science Guy" Nye.

If you had a friend whose career plans involved being an Internet visionary, a speechwriter and musician, you would probably tell them not to quit their day job. Michael Hawley was remarkable precisely because it is not only unusual for anyone to make lasting achievements in one area, it is almost vanishingly rare for someone to be truly accomplished in multiple disparate fields. In Peter Doherty's words, "bright people who hop around from one topic to another often achieve very little". When considering the state of your creative research, be aware that it is often more fun to start something new than it is to persist with a project until it is truly complete.[3] When you and others evaluate the overall impact of your work, however, it is the work that gets finished that counts, not the number or variety of things that you started.

Doherty's advice doesn't mean that you need to spend the next 30 years of your life working on the same esoteric sub-sub-sub-topic in your field. It is a trenchant reminder, however, that the time scales required to make significant progress on important problems are long (see Chapter 1 for more discussion of this point). Be realistic and patient in your expectations with what can be achieved in the time available to you and be willing to continue your work for years in order to accomplish great things.

Be Generous and Culturally Aware

If all work in your field was performed on the planet Vulcan from the Star Trek universe, the success or failure of your research would be judged by Mr. Spock and his fellow Vulcans in purely logical terms. As you have no doubt experienced, this is not how our world actually works. The ways we relate to and are perceived by others in our professional spheres can have long-lasting implications for your career success. As Peter Doherty wrote, "when it comes to awards and prizes, it is probably important to have as few enemies as possible". This sentiment also applies to many of the competitive decisions that are necessary at less rarefied levels, including funding decisions on grant proposals, invitations to speak at conferences and professional promotions.

---

[3] By "complete" I mean reaching a stage where a useful contribution to the field has been made that can be reported in a paper or some analogous output. The nature of research is that it is extremely unusual for a project to be complete in the sense that nothing further could be added or learned in the future.

How can you go about making friends and avoiding making enemies? I hope that some avenues are already obvious: don't be dismissive or condescending to others, listen respectfully when others are speaking and so on. This doesn't mean that you have to automatically set aside your opinions and let others have their way. As one of my colleagues likes to point out, a useful definition of diplomacy is being able to tell someone to go to hell in such a way that they look forward to the trip.

Giving a talk or writing a paper presents you with many opportunities to be generous to others. A negative illustration of this idea was given in Chapter 6 as Mistake #5 for giving bad presentations ("ignore the contributions of others"). Doherty articulated this important idea in a positive way by saying "freely acknowledging the achievements of others is a sure sign of someone who is confident in their own worth and integrity". As you work on writing papers about your work or preparing talks, make giving credit to others an explicit step in your efforts.

The discussion above focuses mainly on people who are "above" you in the research hierarchy in the sense that they are reviewing your manuscripts or making judgments about hiring or promoting you. Viewed in this hierarchical way, there are also many people who are "below" you, for instance, lab technicians, administrative staff and your building's cleaners, for example. Hopefully, you already know that you should be friendly and respectful of all these people, but let me give you a pragmatic reason why doing so is in your self-interest. It is extremely likely that at some stage in your work, you will have an "emergency" situation where you need someone else to drop what they are doing and help you immediately. Perhaps you will need help to finish the paperwork for a crucial grant proposal on a short deadline or fix some vital piece of lab equipment. If you have developed a reservoir of good will and respect over a long period of time with the person you are asking for help, they are likely to be genuinely delighted to help you. If instead you have treated the person poorly in the past, you might find that they don't see your request to be quite as urgent as you might like.

Tell the Truth

Building a positive reputation as a researcher takes a long time. A hard-earned positive reputation can be lost almost overnight if your truthfulness is questioned. As Doherty put it, "Telling the truth about data is an absolute requirement". Being untruthful about data can take many forms, from omitting information that might imply negative results to outright fabrication of data. The story of Elizabeth Holmes and the rise and fall of Theranos in Chapter 1 is one example of the extreme consequences that can flow from people not being truthful about their research.

Being scrupulously truthful about your work doesn't mean that mistakes will never happen. A key feature of scientific progress is that ideas that were once thought to be true are shown to be false. Consider the situation of the physicists mentioned above who announced in 2011 the measurement of neutrinos traveling faster than the speed of light. Instead of digging in their heels and dogmatically insisting their experiment was correct, they performed additional checks and communicated openly with other experimental teams. Once additional experiments found that neutrinos traveled at the speeds expected from known physics and that the original measurements were in error, the team published a detailed paper clarifying their results. I suspect that this approach enhanced rather than diminished the stature of the scientists involved.

A common example in discussing ethics in business is the so-called "Newspaper Test", which asks how a decision-maker would feel if the details of their decisions and motivations were reported in a local newspaper. A variation of this test is useful for thinking about how to handle data in your own research. If a knowledgeable reviewer of the manuscript you sent to a journal or a grant proposal you have written was given unfettered access to every stage of your work, from the design of each experiment to the intricacies of your data analysis, would their opinion of your work become higher or lower? There is a Chinese proverb that says "The good anvil does not fear the hammer". Do your work every day in ways that mean you could welcome scrutiny instead of living in fear of what it might uncover.

Be Persistent, But Be Prepared To Fail

Accomplishing impactful research requires taking risks and a tolerance for uncertainty. If the hypotheses you are exploring in your research are simple enough that they can be validated or disproved in a few days, then they almost certainly are not going to be impactful. Andrew Wiles' six-year effort to prove Fermat's Last Theorem (see Chapter 3) may be extreme, but it is a reminder that great research achievements don't happen overnight. One of the reasons that Wiles' story is inspirational, of course, is that he was ultimately successful. But research doesn't always work this way: hypotheses are discarded because new evidence comes to light and ideas for new technologies fade away because competing methods work better or just because the idea can never quite be made to work.

The risk of failure means that a certain level of personal resilience is required of any successful researcher. As Doherty put it, "People who can't deal with failure or acknowledge to themselves that they have been wrong should probably avoid a life based in research". The second part of this description, the ability to acknowledge being wrong, is a prerequisite for making progress in research. When was the last time you were wrong about something in your work, and how did you react when your error was revealed? Hopefully, the incident that comes to mind is something that helped you refine your work and the experience of finding something wrong was a useful step forward. There is often more to learn from errors and failures than when everything goes smoothly. It can be useful after the fact to analyze a failure in a research project to seek patterns that could be avoided in the future. The same principle applies as you learn about work by other researchers, so be attuned while reading papers and listening to talks to clues about not only went right but what went wrong in other studies.

Take Care of Yourself and Live a Long Life

This book began in Chapter 1 by looking at the stamina that is required to have success in research and similar creative fields. Truly impactful research takes place over periods of years, not days or weeks. Rewards for extraordinary research often lag many years behind the original work being done. To use a sporting analogy, making a career of research is an ultramarathon, not a 100 meter sprint.

To be able perform creative work at a high level for periods of years or decades, it is vital to take care of yourself physically and mentally. In Peter Doherty's words, "Good habits start early: eat and drink moderately, take vacations, don't smoke, take regular exercise, avoid extreme sports, seek professional help for suicidal thoughts". His mention of suicidal thoughts points to an especially alarming mental health challenge, but it is important to realize that mental health encompasses a wide range of conditions that affect the lives of an enormous number of people. According to the National Institutes of Health, 18.9% of all adults in the USA suffered from some kind of mental illness in 2017, with an even higher proportion (25.8%) among adults from ages 18–25. Almost a quarter of these individuals had symptoms that "substantially interfere with or limit one or more major life activities". If you are concerned that you are suffering from a mental health challenge, *do not feel that you are alone*. Many other people have faced similar challenges, including many who have gone on to have long, productive and satisfying careers in research.

If you broke your leg in an accident of some kind, you would visit a doctor to get treatment and use crutches or other aids while you recovered. It is also possible that a doctor's visit could tell you that your leg isn't broken and that it will feel better after some rest. It is helpful to view mental health challenges in a similar way: seek professional diagnosis and follow the guidance of experts who can give you context and strategies for coping with whatever challenges are causing you distress.

Enormous numbers of people around the world live in conditions of financial insecurity or work in jobs that are "just a paycheck". If you are able to forge a career that not only provides for you financially but also brings you genuine satisfaction, then you are fortunate indeed. But your career success does not define your worth as a person. Even during days (or months) when your work isn't going the way you would like it, you can choose to interact positively with the people around you and find satisfaction in activities and relationships outside of your work.

Your Time Is Precious

Regardless of the brilliance of your insight or your dedication to your work, there are only 24 hours in a day and seven days in a week. The issue of time management has come up multiple times in this book, especially in Chapter 2's discussion of deep work. It is not practical to fill your entire schedule with deep work, but the issue of doing too much deep work is not something that most of us ever have to worry about. Instead, the demands and distractions of everything else (shallow work and frippery, in the language from Chapter 2) persistently threaten to squeeze deep work out of your daily calendar. Doherty advises, and I agree, that it is best to view time as a precious resource.

If you bumped into a colleague in a corridor and they asked you for a few percent of your weekly salary, they would probably have to make a very compelling case you for to agree. Now imagine that the same colleague asked you to do something with them for an hour at a time when you were planning on some focused work. Unless your capacity for deep work is impressively high, that hour can easily represent more than a few percent of your deep work for a week. This point of view doesn't mean you should always say no to your colleagues' requests, but it illustrates the precious nature of your time.

The example above focuses on managing your time on a daily or weekly time scale. As your career develops, you will also need to make decisions about longer term demands on your time. Doherty has some pointed advice on maximizing the impact of your research: "Avoid prestigious administrative posts". Setting aside the idea that many researchers might think that "prestigious" and "administrative" are oxymorons, let's unpack Doherty's idea. Doing deep, impactful research takes sustained effort over long periods of time. Taking on any role that will fill significant chunks of your calendar with meetings or other activities makes maintaining a focus on your research goals more difficult. As I write this, I have been head of a large academic department for a number of years, so in a sense I have ignored Doherty's advice. I have experienced firsthand, however, the significant satisfaction that can come from helping a broad cross-section of students, researchers and colleagues be more successful in their work. As I mentioned at the beginning of the chapter, there are many ways to make a positive impact on the world.

When an opportunity to lead an initiative or program comes your way, how should you decide what to do? First, recognize that you can say no. Gracefully and promptly saying no to a request for your time rarely leads to ill will. Second, examine your motivations. Does the opportunity lie in something that you have benefitted from greatly in the past, or something you have long thought is important but could be done better? These are good reasons to consider saying yes. Are you mainly leaning towards saying yes because of guilt or because you think the role will bring you some kind of prestige? These are excellent reasons to say no.

Your time *is* precious, so as much as you can control it, use it towards activities that you will care about in the long run. If you develop habits that lay a foundation for long-term creativity, you will develop the capacity to solve significant problems and derive great personal satisfaction from doing so. It is my hope that the topics you have learned about in this book will lead to actual changes in your work life and that these will help you to, as the book's title says, amaze your friends and surprise yourself.

Chapter Summary

There is no simple formula that can lead to winning the Nobel Prize or reaching similar stellar levels of achievement. There are, however, habits and attitudes that can point you towards long-term success in your creative endeavors. The ten ideas from Peter Doherty listed below are all positive traits to aspire to in your work.

- *Try to Solve Major Problems and Make Big Discoveries*: take time to understand what the important challenges in your field are, and do your best to focus your work on topics that will help resolve these challenges.

- *Acquire the Basic Skills and Work with the Right People*: be conscious of the need to develop core skills in your work and be willing to invest substantial amounts of time to this end.

- *Talk About the Problem*: boost your creativity by talking openly about roadblocks in your work with people from diverse fields and backgrounds and by learning about their work.

- *Learn to Write Clearly and Concisely*: skillful communication can make your career soar, so take time to deliver your results in clear and memorable forms.

- *Focus, Don't Be a Dilettante*: solving important problems inevitably requires long periods of time.

- *Be Generous and Culturally Aware*: personal relationships and impressions matter, so treat everyone you interact with respectfully and with good humor.
- *Tell the Truth*: scientific communities are highly tolerant of mistakes that get corrected over time, but the same can't be said for lapses in integrity.
- *Be Persistent, But Be Prepared To Fail*: be mentally resilient when your research isn't going smoothly, remembering that the path to significant results is never a smooth one.
- *Take Care of Yourself and Live a Long Life*: make long-term investments of time and habits to maintain your physical and mental health and seek professional help if health issues impede your enjoyment of life.
- *Your Time Is Precious*: everybody has the same amount of time available each day. Make conscious choices about how you spend your time and don't be afraid to say "No" to requests that will take time away from more important activities.

Further Resources for Chapter 7

This chapter expands on ideas described by Peter Doherty in his wonderful book *The Beginner's Guide to Winning the Nobel Prize* (Columbia University Press, 2006). This book also touches on other interesting topics in science policy and gives a personal history of the work that led to Doherty winning a Nobel Prize.

Making a long-term impact with your work often involves taking on leadership roles in a formal or informal way. An excellent resource for thinking about the many issues that come up in leadership is *Leadership by Engineers and Scientists: Professional Skills Needed to Succeed in a Changing World*, Dennis Hess (Wiley-AIChE, 2018). Instead of looking at abstract theories of leadership, this book gives practical advice to people trained as researchers who find themselves leading teams.

The paper mentioned in the discussion of concise writing is Can Apparent Superluminal Neutrino Speeds be Explained as a Quantum Weak Measurement? M. V. Berry, N. Brunner, S. Popescu and P. Shukla, *Journal of Physics*, A 44 (2011) 492001. A paper describing what went wrong in the original experimental measurements is Measurement of the Neutrino Velocity with the OPERA Detector in the CNGS Beam, T. Adam et al., *Journal of High Energy Physics*, 10

(2012) 093. As is common in the world of experimental high energy physics, this paper has more than 180 authors.

Benjamin Franklin's extraordinary life is the subject of Walter Isaacson's biography *Benjamin Franklin: An American Life* (Simon and Schuster, 2004). A brief account of Franklin's methods for self-improvement as a writer by Charles Chu is available from https://qz.com/914305/ benjamin-franklin-wasnt-a-natural-talent-heres-how-he-taught-himself-to-write/

Michael Hawley's life was described by an obituary in *The New York Time* on June 24, 2020: https://www.nytimes.com/2020/06/24/technology/michael-hawley-dead.html. The article points out that Steve Job's widely quoted line "Stay hungry, stay foolish" was written for him by Hawley.

The data mentioned on the prevalence of mental illness in the USA is from the NIH National Institute for Mental Health (NIMH): https:// www.nimh.nih.gov/health/statistics/mental-illness.shtml. NIMH is an excellent resource if you or someone you know is experiencing challenges associated with mental illness.

Index

Printed in the United States
By Bookmasters